Alexiei Dingli

Knowledge Annotation: Making Implicit Knowledge Explicit

Intelligent Systems Reference Library, Volume 16

Editors-in-Chief

Prof. Janusz Kacprzyk
Systems Research Institute
Polish Academy of Sciences
ul. Newelska 6
01-447 Warsaw
Poland
E-mail: kacprzyk@ibspan.waw.pl

Prof. Lakhmi C. Jain
University of South Australia
Adelaide
Mawson Lakes Campus
South Australia 5095
Australia
E-mail: Lakhmi.jain@unisa.edu.au

Further volumes of this series can be found on our
homepage: springer.com

Vol. 1. Christine L. Mumford and Lakhmi C. Jain (Eds.)
*Computational Intelligence: Collaboration, Fusion
and Emergence,* 2009
ISBN 978-3-642-01798-8

Vol. 2. Yuehui Chen and Ajith Abraham
*Tree-Structure Based Hybrid
Computational Intelligence,* 2009
ISBN 978-3-642-04738-1

Vol. 3. Anthony Finn and Steve Scheding
*Developments and Challenges for
Autonomous Unmanned Vehicles,* 2010
ISBN 978-3-642-10703-0

Vol. 4. Lakhmi C. Jain and Chee Peng Lim (Eds.)
*Handbook on Decision Making: Techniques
and Applications,* 2010
ISBN 978-3-642-13638-2

Vol. 5. George A. Anastassiou
Intelligent Mathematics: Computational Analysis, 2010
ISBN 978-3-642-17097-3

Vol. 6. Ludmila Dymowa
Soft Computing in Economics and Finance, 2011
ISBN 978-3-642-17718-7

Vol. 7. Gerasimos G. Rigatos
Modelling and Control for Intelligent Industrial Systems,
2011
ISBN 978-3-642-17874-0

Vol. 8. Edward H.Y. Lim, James N.K. Liu, and
Raymond S.T. Lee
*Knowledge Seeker – Ontology Modelling for Information
Search and Management,* 2011
ISBN 978-3-642-17915-0

Vol. 9. Menahem Friedman and Abraham Kandel
Calculus Light, 2011
ISBN 978-3-642-17847-4

Vol. 10. Andreas Tolk and Lakhmi C. Jain
Intelligence-Based Systems Engineering, 2011
ISBN 978-3-642-17930-3

Vol. 11. Samuli Niiranen and Andre Ribeiro (Eds.)
Information Processing and Biological Systems, 2011
ISBN 978-3-642-19620-1

Vol. 12. Florin Gorunescu
Data Mining, 2011
ISBN 978-3-642-19720-8

Vol. 13. Witold Pedrycz and Shyi-Ming Chen (Eds.)
Granular Computing and Intelligent Systems, 2011
ISBN 978-3-642-19819-9

Vol. 14. George A. Anastassiou and Oktay Duman
*Towards Intelligent Modeling: Statistical Approximation
Theory,* 2011
ISBN 978-3-642-19825-0

Vol. 15. Antonino Freno and Edmondo Trentin
Hybrid Random Fields, 2011
ISBN 978-3-642-20307-7

Vol. 16. Alexiei Dingli
*Knowledge Annotation: Making Implicit Knowledge
Explicit,* 2011
ISBN 978-3-642-20322-0

Alexiei Dingli

Knowledge Annotation: Making Implicit Knowledge Explicit

 Springer

Dr. Alexiei Dingli
Department of Intelligent Computer Systems,
Faculty of Information and Communication Technology,
University of Malta,
Msida MSD 2080
Malta
E-mail: alexiei.dingli@um.edu.mt

ISBN 978-3-642-26819-9   ISBN 978-3-642-20323-7 (eBook)

DOI 10.1007/978-3-642-20323-7

Intelligent Systems Reference Library   ISSN 1868-4394

Typeset & Cover Design: Scientific Publishing Services Pvt. Ltd., Chennai, India.

Printed on acid-free paper

9 8 7 6 5 4 3 2 1

springer.com

I would like to dedicate this book to my dear children Ben and Jake, my wife Anna, our parents and the rest of our family. I would like to thank God for being there when things went wrong and for opening new doors when I found closed ones.

Preface

If we want to create the web of the future, an absolute must is to address the issues centred around creating and harvesting annotations. In essence, this future web is not a parallel web but rather a metamorphosis of the existing web. Basically, it needs to tackle two main issues, the first is to have rich websites designed for human consumption and simultaneously, it also needs to offer a representation of the same content which can be digested by software programs.

Unfortunately, we feel that the literature which exists on this subject is limited and fragmented. Now that the study of the web has been consolidated in a field known as Web Science we need to reorganise our thoughts in order to move forward to the next phase. Properties of the web such as redundancy, will gain more and more importance in the coming years so it is imperative to make people aware about them in order to help them create new techniques aimed at exploiting them.

In synthesis, our aim behind this document is to interest a general audience. Unfortunately, since few people are yet aware of the science behind the web and its problems, more expository information is required. So far, the web has been like a huge elephant where people from different disciplines look at it from different perspectives and reach varied conclusions. Until people understand what the web is all about and its grounding in annotation, people cannot start appreciating it and until they do so, they cannot start creating the web of the future.

January 2011

Valletta, Malta
Alexiei Dingli

Acknowledgements

It would be incredibly difficult to list here all the people that helped me throughout these past months to complete this work, so I would like to thank all those who made this piece of work possible. Those who had to bear with my tempers when things seemed impossible to achieve. Those who were next to me when the weather got cold and typing on the keyboard became an incredible feat. In particular I wanted to thank Professor Yorick Wilks, an icon in the field of Artificial Intelligence; the person who believed in me and who gave me the opportunity to become what I am today. My supervisor, my colleague, my friend and to some extent my extended family. I can never thank him enough for his unfailing support. Finally, I just wanted to say a big thank you to everyone who made this document possible, a small step towards the realisation of my dream.

Contents

List of Figures

List of Tables

Acronyms

AAL	Ambient Assisted Living
AJAX	Asynchronous JavaScript and XML
AR	Augmented Reality
BBS	Bulletin Board Services
CAPTCHA	Completely Automated Public Turing test to tell Computers and Humans Apart
CMU	Carnegie Mellon University
COP	Communities of Practise
CSS	Cascading Style Sheets
DAML	DARPA Agent Markup Language
DARPA	Defense Advanced Research Projects Agency
ESP	Extra Sensory Perception
EU	European Union
GIS	Geographical Information System
GML	Generalised Markup Language
GUI	Graphical User Interface
GWAP	Games With A Purpose
HTML	HyperText Markup Language
HTL	Human Language Technologies
IBM	International Business Machines
IE	Information Extraction
II	Information Integration
IM	Instant Messaging
IP	Internet Protocol
IR	Information Retrieval
IST	Information Society Technologies
ML	Machine Learning
MUD	Multi User Dungeons
NLP	Natural Language Processing
OCR	Optical Character Recogniser
OIL	Ontology Inference Layer

OWL	Web Ontology Language
P2P	Peer-to-Peer
POI	Points of Interest
RDF	Resource Description Framework
RDFS	Resource Description Framework Schema
RFID	Radio Frequency Identification
RSS	Really Simple Syndication
SGML	Standard Generalised Markup Language
SOAP	Seal Of APproval
SW	Semantic Web
UK	United Kingdom
URI	Unified Resource Identifier
URL	Unified Resource Locator
US	United States of America
VOIP	Voice over IP
W3C	World Wide Web Consortium
WML	Wireless Markup Language
WWW	World Wide Web
WYSIWYG	What You See Is What You Get
XHTML	Extensible HyperText Markup Language
XLink	XML Linking Language
XPointer	XML Pointer Language
XML	Extensible Markup Language

Part I
A World of Annotations

"When patterns are broken,
new worlds can emerge."

Tuli Kupferberg

Chapter 1
Introducing Annotation

Annotation is generally referred to as being the process of adding notes to a text or diagram giving explanation or comment. At least, this is the standard definition found in the Oxford Dictionary [75]. As a definition, it is correct but think a little bit about today's world. A world where the distinction between the virtual and the real world is slowly disappearing; physical windows which allow people in the real world and people in a virtual world to see each other (such as the virtual foyer in [141]) are starting to appear. These portals are not only limited to buildings, in fact the majority of them find their way in people's pockets in the form of a mobile phone. Such phones go beyond the traditional voice conversations and allow their users to have video calls, internet access and the list of features can go on to even include emerging technologies such as augmented reality [223]. This extension of reality is obviously bringing about new forms of media and with it, new annotation needs ranging from the annotation of videos [199] or music [217] for semantic searches up to the annotation of buildings [192] or even humans [162]. A better definition of annotation can be found in the site of the World Wide Web Consortium (W3C) Annotea project[1] which states that:

> By annotations we mean comments, notes, explanations, or other types of external remarks that can be attached to any Web document or a selected part of the document without actually needing to touch the document.

Obviously even though this definition is far better than the previous one, we still need to handle it cautiously because it also opens a new can of worms. There is still an open debate about the issue of whether annotations should be stored within a document or remotely (as it is being suggested by the Annotea team). But before divulging further into this issue, the following section will focus on why there is this need to create annotations.

---

[1] http://www.w3.org/2001/Annotea/

A. Dingli: Knowledge Annotation: Making Implicit Knowledge Explicit, ISRL 16, pp. 3–17.
springerlink.com

1.1 Physical Annotations

From an early age, children start the game of labelling objects. In fact, it is very common to see a child point at something and repeating its name over and over again. This is very normal since the child is trying to organise and process his thoughts to convey a message. Although this might seem simple, in reality it is much more complex since as [225] have shown, children are not only labelling objects but also creating a hierarchy of labels which describes the world. When we grow older, this labelling process becomes more discrete and automatic. We tend to do it in our heads without even realising that we're doing it. However, this process resurfaces and becomes annotation when we handle printed media.

Did you ever read a book or any form of printed material and felt the need to scribbled something on the book's text or margins? If you did, you just annotated a text. [127] goes through the various aspects of annotations in books and also studies the rational behind them. A known fact (which is reinforced in her findings) is that when we are still young, we are discouraged by our guardians or educators to scribble on books for the fear of ruining them. However this goes against popular wisdom. In fact according to [86], Erasmus used to instruct his students on note taking in order to prepare themselves for their speeches. In his address, he tells them to create special markers in order to differentiate specific sections in the text.

As we grow older, we tend to let go of this fear and find it convenient to scribble on the document itself. However, this only holds if we own the document (and generally only if the document does not have some intrinsic value) or if we are asked to add annotations on the document. [127] considers annotation as being a monologue between the annotator and either his inner self or the author. In fact, she noticed that in general, annotators mark parts of the document which they might need to reuse at a later stage, as a form of self note (as suggested by Erasmus earlier). The reason why annotations are inserted in texts differs from one person to another, even on the same document. A chef might scribble on a recipe book to insert additional ingredients such as meat to a particular recipe. The editor of the same book might add comments on the layout of the recipe or some corrections. Annotations could also have a temporal dimension. A comment written today might not be valid tomorrow. If the restaurant where the chef works decided to offer vegetarian alternatives, the previous annotations (pertaining to meat) would have to be removed. Annotations could be just personal thoughts added to a document or they could be created to share content with someone else. Sticky notes are a popular way of adding additional information to physical objects; you can stick them into documents or onto the object they apply to (a package) or just put them in the front of the fridge for everyone to read. However, it was also interesting to notice that annotators frequently left comments to the author of the document. They do so with the consciousness that the author will probably never read their comments and this gives them that additional intimacy of expressing themselves. According to [183], reading is not a dialogue between the reader and the author but rather an expression from the text to the reader. The messages sent from the text permeate the reader's thoughts and are trapped within the reader's mind where they can be nurtured or pruned. All of this is within the control of the reader and eventually, some of these thoughts result in annotations which change the document forever. As long as the document is kept by the reader who annotated it, the thoughts are not disclosed.

However, when the annotated documents are circulated to others, the story changes. The effect of the annotations can vary depending on the reader and it might raise different emotions. This issue is obviously accentuated when we deal with digital documents.

1.2 Digital Annotations

With digital documents, the annotation process becomes much easier. Most of what we already discussed with physical documents still holds. Readers still annotate for the same reasons mentioned earlier, however, me must also add to this other aspects.

The origins of digital annotations date back to the 60s when International Business Machines (IBM) embarked on a project [104] whose result was the creation of the Generalised Markup Language (GML). Originally, it was only intended as a data representation system targeting legal documents. However, IBM saw other uses to this language and today, it forms the basis of most markup languages (Standard Generalised Markup Language (SGML), Extensible Markup Language (XML), Extensible HyperText Markup Language (XHTML), Wireless Markup Language (WML), etc) and its applications range from defining semantics to specifying layouts. The same period also saw the conception of another important concept for digital annotations, the idea of HyperText. The term was originally coined by Ted Nelson and refers to the concept of text, having references to other texts which can be followed by simply clicking a mouse. This concept is especially important for external annotations. In the 70s, the TEX typesetting system was created by Donald Knuth. Thanks to such a system, for many years, in-line publishing commands similar to annotations were the most common way of formatting documents (in tools such as Latex, changing the layout was simply a matter of annotating the text with the appropriate command followed by curly brackets such as \textbf{**words to be in bold**}). It promoted the idea that layout and content can be mixed in the same document. In fact, the use of annotation was boosted further with the creation of HyperText Markup Language (HTML) where web documents contain both the information and its layout in the same document.

Annotations have been around for decades and people have been using them to record anything they like. The type of annotation used varies between different programs however Figure 1.1 gives a summary of the most popular annotation types.

Lorem ipsum dolor sit amet, consectetur adipiscing elit. Cras mollis commodo faucibus. Vestibulum arcu metus, egestas quis mollis sed, egestas sit amet arcu. Donec gravida ipsum sit amet orci sollicitudin feugiat. Morbi leo sapien, feugiat sed laoreet id, ullamcorper non nunc. Vestibulum non ligula risus. Suspendisse feugiat felis a mauris lacinia vitae aliquam odio scelerisque. Phasellus ultrices egestas interdum. Nunc accumsan, diam id volutpat condimentum, metus mi cursus odio, sit amet hendrerit ipsum est vitae nunc. Cras vulputate, purus vel pretium mollis, risus purus pharetra metus, a adipiscing mauris dolor ultrices ipsum. Suspendisse eget ligula a ligula congue tempus ut condimentum diam. Cras sit amet mauris tortor, id bibendum nibh.

More research on this ...

Lorem ipsum dolor sit amet, consectetur adipiscing elit. Cras mollis commodo faucibus. Vestibulum arcu metus, egestas quis mollis sed, egestas sit amet arcu. Donec gravida ipsum sit amet orci sollicitudin feugiat. Morbi leo sapien, feugiat sed laoreet id, ullamcorper non nunc. Vestibulum non ligula risus. Suspendisse feugiat felis a mauris lacinia vitae aliquam odio scelerisque. Phasellus ultrices egestas interdum. Nunc accumsan, diam id volutpat condimentum, metus mi cursus odio, sit amet hendrerit ipsum est vitae nunc. Cras vulputate, purus vel pretium mollis, risus purus pharetra metus, a adipiscing mauris dolor ultrices ipsum. Suspendisse eget

ligula a ligula congue tempus ut condimentum diam. Cras sit amet mauris tortor, id bibendum nibh.

Lorem ipsum dolor sit amet, consectetur adipiscing elit. Cras mollis commodo

faucibus. Vestibulum arcu metus, egestas quis mollis sed, egestas sit amet arcu.

Fig. 1.1 A document showing various forms of annotations

Textual Annotations are various and most of them find their origins in word proces-
sors. These annotations include amongst others underlined text , ~~strike through text~~ and highlighted text. The colour of the annotation is also an indication.
Most of the time, it is related to a particular concept thus an annotation is always
within a context. More complex annotations such as highlighting multiple lines
were also inserted thus offering more flexibility to the user. The highlight essentially is a way of selecting an area of text rather than just a few words thus giving
more power to the annotator. The power is derived from the fact that the annotation is not limited to the rules of the document since it can also span multiple
sentences and also highlight partial ones.

Vector Annotations are more recent. Their origin is more related to graphical packages even though modern annotation editors manage to use them with text. Vector
annotation is made up of a set of lines stuck together generally denoting an area
in the document. The shape of the line varies from the traditional geometrical
figures such as circles, squares, etc to freehand drawing. The latter is obviously
more powerful and very much adapted to graphical objects. In fact, in Figure 1.1,
we can see that the bear has been annotated with a freehand drawing. This means
that if someone clicks on the bear, he is taken somewhere else or information
related just to the bear is displayed. To identify objects in images, such as the
face of a person, more traditional shapes can be used such as the square in the
case of Figure 1.1.

Callout Annotations take the form of bubbles or clouds and are normally used to
provide supplementary information whose scope is to enrich the document's content. In the case of Figure 1.1, the red callout is used as a note to the annotator
whereas the cloud is used to express the thoughts of the baby. The use of these
annotations are various; there's an entertaining aspect where they are used to
highlight the thoughts or discourse of the people involved. They are also very
useful when it comes to editing documents especially during collaborative edits
where the thoughts of the editors can be shared and seen by others.

Temporal Annotations take the forms mentioned earlier however they are bound
by some temporal restriction. These annotations are mainly used in continuous
media like movies or music where an annotation can begin at a specified period
of time and last for a predefined period.

Multidimensional Annotations can take the forms mentioned earlier however
rather than having just an (X,Y) coordinate to anchor them to a document, they
also have other dimensions such as (X,Y,Z) in order to attach them to 3D objects.
These annotations might also have a temporal dimension such as in the case of
3D movies. With multidimensional datasets, annotation is also possible however
they are much more difficult to visualise graphically.

With all these different annotation tools, it is also important to understand why people annotate digital documents. There are various reasons for this; first and foremost because it improves the web experience and secondly because digital documents are easier to modify and distribute.

The improvement to the web experience might not be immediately evident however if we delve into the usages of annotations, this will definitely become obvious.

1.2.1 Annotations Helping the Users

Irrespective of the medium being use, it is important that annotations are inserted to supplement the document being viewed. This might include further explanations, links to related documents, better navigational cues, adding interactivity, animating the document, etc. The annotations should not reduce the quality of the document or distract the user with unrelated stuff.

The application used to view the annotation (be it a browser, a web-application, etc) should be careful about the invasiveness of annotations. Let's not forget that a user should be made aware that an annotation of some sort exists yet, this must occur in a discrete way which allows the user to tune the invasiveness of such an annotation. This kind of problem can be particularly observed when dealing with video annotations, some of these annotations occupy parts of the viewpoint in such a way that the video is barely visible. This definitely defies the scope of having annotations.

When annotations are created, they also have a contextual relationship. The object they annotate being a piece of text, a 3D model or any other object has some sort of link to the annotation. Because of this, users expect the annotations to be relevant and give value to the document. However, annotations are sometimes used for other purposes such as advertisements, subscriptions, voting, etc. People tend to be particularly annoyed with these kind of annotations and tend to see them as another form of spam. The reason for this has to do with closure as explained in [80]. When people go online, they normally do it to reach a particular objective which might range from watching a movie to learning about quantum mechanics. This particular objective is normally made up of various subgoals and each one of them has a start and an end. Closure occurs each time a subgoal reaches the end. Whilst working to achieve the subgoal, users get very annoyed as can be seen in [201] if they are interrupted with something which is unrelated or which does not help them reach the end of their goal. The study also shows that if the interruption is made up of somewhat related information, users are more prone to accept it.

Even though annotations are mainly inserted for the benefit of other users, in some cases, annotations even give a financial return to the creator of the annotation. This is achieved using two approaches:

Fig. 1.2 A movie on YouTube (See http://www.youtube.com/watch?v=TnzFRV1LwIo) with an advert superimposed on the video

Fig. 1.3 A clear reference to a particular mobile phone brand in the movie Bride Wars (2009)

The direct approach involves selling something directly through the annotation. The annotation might be an small advert superimposed on the video. As can be seen in Figure 1.2, whilst the video is running a semitransparent advert pops up on the screen which allows the user to click on the advert and go directly to the conversion page. This should not be mistaken with the links which will be mentioned in the next section since these direct approaches push the user towards effecting a purchase. Since these annotations are non-intrusive, they are quite popular in various mediums such as in traditional computers, but also in mobile phones having small displays.

Another form of annotation which is somewhat more discrete is the insertion of hotspots into pictures or movies. A hotspot is a clickable area on screen which links directly to another place somewhere else online. In the case of pictures, hot-spots are fixed, however when it comes to movies, hotspots have a temporal dimension because a clickable area can only last for a few frames. Figure 1.3 shows a screenshot from the movie Bride Wars (2009). The screenshot shows clearly the mobile phone used by the actress in the movie but in effect, this is just a discrete promotion for the brand. With the advent of interactive TV such as Joost[2] and Google TV[3] the person watching this movie can simply click on the mobile phone shown in Figure 1.3 and he is immediately taken to the marketplace from where he can purchase the product. This approach offers various advantages over traditional advertising. First of all it is non-invasive since the product is displayed discretely and ties very much with the storyline of the movie. Secondly, it will change the whole concept of having adverts. In traditional settings, a movie is interrupted by adverts whilst in this context, there's no need of interruptions since the movie and the adverts are fused together. This is solely achieved by annotating movies with adverts.

The indirect approach whereby annotations are just links which drive traffic to a particular website and as a result of that traffic, the owner of the website earns money. When people access a particular document and click on an annotation link, it simply takes the viewer to another document. Often what these people would do is then link back to the original document again with another annotation. The notion of having bidirectional links (i.e. links which take you somewhere but which can also take you back from where you originally left) is not new and in fact it was one of the original ideas of Professor Ted Nelson for the web. In these links, one side of the link can be considered the source and the other side, the target. This idea is very different from the Back button found in a web browsers because that button has nothing to do with HTML but it is essentially a feature of the browser. In fact, there is nothing inherent in HTML links that supports such a feature since the target of a link knows nothing about its source. This has been rectified with the XML Linking Language (XLink)[4] standard and with the creation of techniques such as linkbacks which allows authors to obtain notifications of linkages to their documents. By providing ways of taking the user back, the system guarantees closure since the user's goals are not interrupted whilst traffic is being diverted to another website. Once a website has traffic, it adopts various approaches to make money such as advertising, direct sales, etc.

Another indirect approach is normally referred to as a choose-your-own-adventure-story style. This approach became famous decades ago in the publishing industry where rather than reading a book from cover to cover, the user reads the book page by page. At the end of every page, the reader is asked to make a choice and depending on the choice made, he is then instructed to

---

[2] http://http://www.joost.com/

[3] http://www.google.com/tv/

[4] http://www.w3.org/TR/xlink/

continue reading a particular page number. Thus, the flow of the book is not linear but appears sporadic. By using this approach, each user is capable of exploring various interwinding stories in the book having different endings. With the advent of HTML and the versatility of links, this approach was quickly adapted to web pages. It also took a further boost with the introduction of multimedia. In fact, what happens is that advertisers are using this approach to create a story which spans different forms of media (text, videos, etc.). The story is written for entertaining purposes however it conceives within it some subliminal messages. A typical example is the "Follow your INSTINCT" story on YouTube. Essentially, this story was created by a phone manufacturer to promote a mobile phone. However, the phone is just a device which is used by the actors in the story. To make it engaging, at the end of the movie, the creatures use different annotations to give the user a choice. This then leads the user to other movies having different annotation and various other choices. At no point is the product marketed directly and in fact, there are no links to the mobile phone manufacturer in the movie, however the flow of the story is helping the viewer experience the potential of the product being sold.

1.2.2 Modifying and Distributing Annotations

By nature, digital documents are easier to modify and distribute. According to [91], annotations can have two forms embedded or external. Embedded annotations are stored within the same documents (such as HTML[5]). The positive aspect of such an approach over physical documents is that a large number of annotations can be added to the same document. Even though they are added, with contrast to physical documents, they can also be removed (if the annotations are inserted with proper tags which distinguish them from the main text) without necessarily altering the original document forever. The downside of these kind of annotations is that the annotator definitely needs to have the ownership rights of the document. Without these rights, no modifications can be added to the document. When it comes to external annotations, annotators can still add a large number of levels of annotations. Since the annotations are stored somewhere external to the original text, the original text is preserved as it was intended by the author. The last advantage over embedded annotations is that annotators do not need the document's ownership to add an annotation since the original document is not being modified. Obviously this approach has its own disadvantages too. External systems must be setup in order to support the annotation process. Since the annotations are not stored directly in the document, if the original document is taken offline, the links between the document and the annotations break, thus resulting in orphan annotations.

With regards of the distribution of digital media, the fact that digital documents can be easily sent through a network makes them ideal for spreading annotations.

---

[5] http://www.w3.org/MarkUp/

However, having a network containing several billion pages[6] does not help. Pages have to struggle to get noticed in the ocean of digital information. According to [109], the best search engine only manages to cover around 76% of the index-able web. Apart from this, there's a deeper web, hidden to search engine. [28] estimates it to be around 500 times larger than what we commonly refer to as the World Wide Web. So depending on whether the document is located in the shallow or deeper web, it will make a huge difference when it comes to sharing annotations. But even if the document manages to make it to the surface web, [147] found that for search engines, not all documents are equal. In fact most popular search engines calculate their rankings based upon the number of links pointing to a particular document. Popular approaches such as [176], which rely on the use of features other than the page content further bias the accessibility of the page. The effect of this is that new qualitative content will find it hard to become accessible thus delaying the widespread of new high-quality information. Also, if the annotations are external to the document, it is very unlikely that the search engine crawlers will manage to locate them and index them accordingly. However, the emergence of a new paradigm called Web 2.0 changed all of this!

1.3   Annotations and Web 2.0

Back in 2004, when O'Reilly Media and MediaLive International outlined the Web 2.0 paradigm (which was later published in [174]), it brought about different reactions. Many researchers started questioning ([236], [103], [128]) this new concept. Some argued that it was just a new buzzword[7], others hailed Web 2.0 as being the beginning of a web revolution[8]. The accepted meaning of Web 2.0 can also be found in Tim O'Reilly's original article [174] where he stated that ...

> the "2.0-ness" is not something new, but rather a fuller realisation of the true potential of the web platform

So essentially, we are not referring to new technologies, in fact, technologies such as Asynchronous JavaScript and XML (AJAX)[9], XML[10], etc have been around for quite some time. But Web 2.0 is all about using these technologies effectively.

As can be seen in [156], annotation in the form of tagging[11], is taking a prominent role in Web 2.0 ([209], [119], [238], [18]) and can be seen as an important feature of several services:

---

[6] http://www.worldwidewebsize.com/

[7] A transcript of a podcast interview of Tim Berners-Lee in 2006
http://www.ibm.com/developerworks/podcast/dwi/cm-int082206.txt

[8] http://www.time.com/time/magazine/article/0,9171,1569514,00.html

[9] http://www.w3.org/TR/XMLHttpRequest/

[10] http://www.w3.org/XML/

[11] The process of attaching machine readable annotations to an object.

delicious.com allows users to tag their bookmarks for later retrieval.

sharedcopy.com makes use of bookmarklets which provide annotation functions to any website.

docs.google.com is an online word processing system having (amongst others) a function to insert colour coded comments within the text.

Facebook.com provides tools for the creation of social tags whereby people are tagged in photos thus allowing the system to create social graphs highlighting relationships between people.

Flickr.com allows the insertion of up to 75 distinct tags to photos and videos. Apart from this, it also allow geotagging.

gmail.com does not uses tags but labels. Essentially they are used in a similar way to tags whereby several labels can be assigned to different emails thus providing quick retrieval.

youTube.com allow users to enhance the content of a video using various annotations.

The list is obviously non exhaustive, however it provides a good representation of typical Web 2.0 applications. A common factor in most of them is that they are not simply bound to text but most of them can also handle pictures, movies and other forms of media. Since multimedia documents are manually annotated, they are easier to index by search engines thus providing a partial solution to the problem of distributing qualitative material over the World Wide Web. When the documents are distributed, so are the annotations and the thoughts of different authors concealed in those annotations.

However, this approach has its own problems. Since annotations are nothing more than words, there is no explicit meaning associated to them and because of this, issues such as homonomy and synonymy arise. To partially solve this problem, some of these systems group the tags into folksonomies. These hierarchies do not provide an exhaustive solution to this problem however, studies by [113] show that eventually, consensus around shared vocabularies does emerge even when there is no centrally controlled vocabulary. This result is not surprising when considering the 8 patterns of Web 2.0 (see [174]). In fact one of these patterns focus on the need for Web 2.0 applications to harness collective intelligence and by leveraging on this collective effort, better annotations can be produced. This idea emerges from [186] where the author states that:

> Given enough eyeballs, all bugs are shallow.

Originally the author was referring to open source software development but it can also apply to collective annotations. Another interesting aspect of Web 2.0 is the principle that "Software is above the level of a single device". What this essentially means is that we should not be limited to the PC platform. New devices are constantly emerging on the global markets; mobile phones, tablets, set-top boxes and the list can go on forever. Its not just a matter of physical form but also of scale; in fact we envisage that one day, these devices will become so small that they will just disappear [106]. Because of this we need to rethink our current processes.

1.4   Annotations Beyond the Web

New devices offer new possibilities, some of which span beyond the traditional World Wide Web (WWW) into the physical world. Two pioneering fields in this respect are Augmented Reality (AR) and Ambient Assisted Living (AAL).

With the advent of camera phones, AR became possible in one's pocket. Essentially, by making use of the camera, the images are displayed on the phone's screen and the software superimposes on them digital information. An example of this can be seen in Dinos[79][68] whereby a virtual mobile city guide is created in order to help people navigate through a city. Figure 1.4 gives a screen shot of the system while it is running. In this example the annotations are superimposed upon the video and serve as virtual cues. In the picture one can notice three types of annotations:

- Points of Interest (POI) are markers identifying interesting locations on a map. They range from famous monuments (like the examples provided in the picture) to utilities such as petrol stations, etc. In Figure 1.4, two blue markers denoting a POI can be seen, one referring to the "Altare Della Patria" and the other to the Colosseum. It is interesting to note that the position of the tag on the screen is determined by the latitudinal and longitudinal position of the tag. These tags are essentially made up of two parts, a square at the top and a textual label underneath it. The square is filled with smaller stars and circles. Stars are a representation of the quality of the attraction as rated by people in social networking sites. Three stars indicate a good attraction which is worth visiting. No stars inform the tourists that the attraction can be skipped. The circles are an indication of the queue length in the attraction. Three circles denote very long queues whereas no circles indicate no queues. This is an interesting feature of Dinos where it manages to combine real world information with virtual navigation. In fact the system has several cameras installed in various locations around the city which are used to measure queue lengths via an automated process. This information is then analysed and presented to the users in the form of red circles. So in the example shown in Figure 1.4, according to the system the Colosseum is more worth visiting than the "Altare Della Patria" because it has a higher rating (indicated by the three stars) and because there are less queues (indicated by the red circle). The position of the square on top of the label is also an indication of direction. In the case of the Colosseum, the square is located to the left of the tag indicating to the tourist that the user has to walk left to find the Colosseum.

- Virtual adverts are indicated by the red tags. These virtual adverts can be placed all over; be it with walls, free standing, floating, etc. They are normally used to indicate a commercial location. These adverts are normally paid by the owners of establishments thus they have a limited lifetime and they don't have ratings.

The lifetime is defined by the amount of money which the owner pays in order to erect the virtual adverts. They do not have a rating system assigned to them because they are dynamic, thus they expire since they are normally used to give out promotional information.

- Virtual graffiti are shown as speech bubbles. The main difference between a virtual graffiti and other types of annotations in the system is that the virtual graffiti are the only kind of annotations inserted directly by users. In actual fact, they've been inserted by friends of the user (where a friend is someone which is known to the user and whose relationship was certified through the use of a social networking site). These graffiti can be seen represented as a green speech bubble where the friend of the user is recommending the attraction. In actual fact, they can be used for anything, i.e. either to share thoughts, comments, etc. They can also be attached to any object and they are shown each and every time a user is in that location.

Even though we've seen this tourist application, in actual fact, the use of AR is very vast (including assembling complex objects [111], training [82], touring [90], medical [84], etc) but an important use, shared by a large number of applications, is to display annotations. [234] shows how such a system can be used to provide information to shoppers while doing their daily errands. By simply pointing the camera to a product, additional information is displayed about that product. A museum system such as [167] can offer a similar experience with regards to its exhibits. So in theory, anything can be virtually tagged and then accessed using AR systems. The advent of the social web is taking this a step further; we have already seen its application in Dinos however [110] is using it to help shoppers by enhancing the shopping experience by adding social content. According to their research, when buying over the internet, most people make use of social content to help them take a decision. Their application makes use of AR to display reviews related to the product being viewed. In essence AR is providing users with a new way of viewing data, a way which is much more natural since it is inserted directly within the current context.

AAL deals with creating smart buildings capable of assisting humans in their day to day needs. The bulk of the research focuses on vulnerable people; such as the elderly and the sick ([235] [140] [23] [187]). In these scenarios, AAL is used to track the movement of both people and physical objects. Various scenarios can be considered such as; people might be kept away from zones where radiation is in progress, the system might check that objects such as scalpels are not misplaced or stolen and it might also double check that a person undergoing a transfusion is given the correct blood type. The scenarios are practically endless and in all of these, a certain degree of annotation is required. The system is not only tracking people and objects but reasoning on them, inferring new knowledge and where necessary annotating the world model. As explained in [77] [76] this is made possible through the creation of a world model of the hospital. Every person and object is being tracked and the system updates their presence in the world model. This information is available on the handheld device of the hospital staff thus providing staff members with realtime information about the situation inside their hospital.

Fig. 1.4 Some examples of augmented reality and annotations in Dinos

This realtime view of the hospital is extremely important. In case of an emergency such as in a fire outbreak, the system calculates in real time the evacuation plan for all the people inside the hospital. People whose life is in danger (such as those trapped in particular areas of the hospital or patients that are bed ridden) are tracked by the system (using information obtained through their Radio Frequency Identification (RFID) and annotated in the world model so that rescuers will have the full picture at hand on their hospital plan. Obviously, reading a plan of a hospital on a small handheld is not ideal, a step further is the use of [162] [163] whereby information is projected on physical surfaces such as papers, walls, tables, etc. By doing so, the flat 2D plan of the hospital is instantly annotated by projecting on it the information obtained through the world model. [161] shows how even humans can be annotated using projected annotations. The idea might sound strange but think about the potential; imagine you're at a doctor's visit. Just by looking at you, the doctor can view information about previous interventions, your hearth rate (using remote bio-sensing[12]), etc. The information would be projected within context, thus your heart rate would appear on your chest, a previous fracture of the leg might have an X-Ray image projected on the effected part. The annotation possibilities are practically endless and only limited by the imagination of researchers.

1.5 Conclusion

In essence, annotation is all about adding value to existent objects without necessarily modifying the object itself. This is a very powerful concept which allows people independently from the owner of the object to add value to the same object. This chapter clarified what is meant by annotation, why it is needed and the motivations behind its usage. The coming chapter, will deal with the Semantic Web and explains why annotation is so fundamental to create such a web.

---

[12] Bio-sensing is the process of transmitting biological information (such as blood pressure) of an individual to a machine.

Chapter 2
Annotation for the Semantic Web

The web has gone a long way since its humble beginnings of being a network aimed at transferring text messages from one place to another. Today's documents are extremely rich, having all sorts of media; text, images, video, music, etc all interlinked together using hyperlinks. Unfortunately, the downside of this proliferation of the web is that users are simply not coping with the myriad of information available [122] [16] [27]. This is normally referred to as the problem of information overload.

This problem grew due to a number of different factors combined together. Coffman in [61] and [62] analysed the rapid expansion of the internet and he found that as a result of more and more people gaining access to the network, new pages containing all sorts of information are being created. This was accentuated further in recent years with the rise of web 2.0 applications whereby more users are converting from being information consumers to information providers. In [174], it is clearly shown that tools such as wikis and blogs are helping people contribute to the web without actually requiring any particular knowledge on how the web works. Another factor according to [178] is that the nature of the information itself (mentioned in Section 1.2) and the creation of sharing methodologies such as Peer-to-Peer (P2P) are making it easier for people to share their digital contents.

Finally, new technologies (such as Voice over IP (VOIP), Really Simple Syndication (RSS), Instant Messaging (IM), etc) are creating new distribution channels and these channels are creating even more information. To make matters worse, a large chunk of information on the web is made up of either free or semi-structured text thus having no overall structure which makes it hard to identify the relationships in the text.

Existent search engines tried to create some sort of order. For many years, the traditional approach was to identify a small set of words which would represent the information being sought and then match it with words extracted from the collection of documents. The document in the collection with most matches would be the top ranking document. A major improvement over this approach was the PageRank algorithm developed in [176]. The major difference in this algorithm was the recognition that documents on the web are interlinked together using hyperlinks. The links were being treated as sort of votes from one site to the other and these

A. Dingli: Knowledge Annotation: Making Implicit Knowledge Explicit, ISRL 16, pp. 19–24.
springerlink.com

"votes" impacted drastically on the ranking of the search results. Although the improvements provided by this approach were very significant, users soon realised that the group of documents returned, still posed a major problem because users would still have to sift through their pages in order to find the piece of information they are seeking. This is extremely frustrating and confirms the findings of various studies such as [147] [172] [204] [64] [131] which state that the majority of users are using the Web as a tool for information or commerce. However, search engines are not returning that information but only potential pages containing that information.

With the advent of Web 2.0 in recent years, several technologies (Social Bookmarking, Social Tagging, etc) too proposed some improvements (see [198] [232]) over traditional search techniques, however according to studies conducted by [121] it is still too early to quantify the real benefit of such approaches.

2.1   The Rise of the Agents

Towards the start of the millennium, [29] proposed an extension to the current web known as the Semantic Web (SW). Back then, it was clear that the proliferation of the web was making the WWW unsustainable by humans alone. This situation also created financial hurdles on organisations, in fact, [173] expected the spending on content management and retrieval software to outpace the overall software market a few years later. Because of this, the extension of the current web was engineered to make it possible for intelligent agents to understand the web, roam freely around it and collect information for the users. To make this possible, a fundamental change is necessary to the documents found on the web; the agents have to understand what's written on them!

If we visualise a small subset of websites present on the web, we soon realise that most of the digital content is aimed for human consumption. These pages are full of animations, movies, music and all sorts of multimedia elements which are incomprehensible by the computer agents. In fact, for these agents, these elements are nothing more than binary numbers. So the idea behind the SW is to add meaning or semantics to documents which agents can understand and act upon. This is achieved by associating semantic annotations to whole or parts of a documents using information obtained from domain ontologies[1] (as described in [29]) thus resulting in documents having annotations which can be interpreted by agents. If these annotations are well-defined, they can be easily shared between the annotator and the users or agents consuming those annotations. In doing so, there would be a clear agreement between the two and any ambiguities removed. So one of the targets of the SW is to create worldwide standards which act upon heterogeneous resources and provide a link between common vocabularies. Semantic annotation goes beyond traditional annotations because apart from targeting human consumption, it is also intended for machine consumption[228], because of this, a key task of this process is to identify relationships and concepts shared within the same document

---

[1] [108] defines an ontology as a formal specification of a shared conceptualisation.

(and possible beyond). For example, consider the semantic annotation on the word "Paris". Since the annotation is related to an ontology, it links "Paris" to the abstract concept of a "City" which in turn links to the instance of the concept of a "Country" called "France". Thus, it is removing any sort of ambiguities which might arise from other connotations (such as "Paris"[2] the movie or "Paris Hilton" the show girl). With ambiguities aside, information retrieval becomes much more accurate according to [226] since it exploits the ontology to make inferences about the data. This approach is so useful that its use is being investigated in various fields ranging from online commerce [38] [214] to genomics [137] [185].

2.2 Ontologies Make the World Go Round

As mentioned earlier, to organise these semantic annotations in a coherent structure, we normally use an ontology. Essentially, an ontology is a large taxonomy categorising a particular domain. It is not expected to cover everything that exists in the world but only a subset. By managing a subset, it is therefore easier to share, distribute and reach an agreement over the concepts used. In the 90s, different organisations used different structures having different formats. For example, both Yahoo![3] and the Open Directory Project[4] used to categorise the web however, even though they were categorising the same data, their structures were not compatible. To tackle these issues, the first task to create the SW was to find a common base language. This eventually became the XML[5], a subset of the SGML meta language which was originally designed to be a free open standard used to exchange all sorts of information. Even though XML is a powerful[6] language, the fact that it is a meta-language does not provide any advanced constructs but only the basic tools to create other markup languages.

Because of this, since 1999 the W3C[7] has been developing the Resource Description Framework (RDF)[8]. The scope behind [39]'s work was to create a language, understandable by web agents and capable of encoding knowledge found on web pages. This language was based on the idea that everything which can be referenced by a Unified Resource Identifier (URI) can be considered as a resource and any resource can have a number of different properties with values. In fact, RDF is based on triples (made up of a Resource, a Property and a Property Value) and these triples makes it possible for RDF to be mapped directly onto graphs [45] [118] (having a Resource and a Property Value as the endpoints of the graph and the property would be the line joining the two endpoints) as can be seen in Figure 2.1. This mapping is

---

[2] http://www.imdb.com/title/tt0869994/

[3] http://www.yahoo.com

[4] http://www.dmoz.org

[5] http://www.w3.org/XML/

[6] http://xml.coverpages.org/xmlApplications.html lists hunders of markup languages created using XML.

[7] http://www.w3.org/

[8] http://www.w3.org/RDF/

Resource	Property Name	Property Value
France	Capital City	Paris

France	Capital City	Paris

Fig. 2.1 An example of a triple; both in table form and also as a graph

very important since RDF does not only provide a structure to the data on the web but it also allows us to apply the power of graph theory on it.

When researchers started using RDF, it was immediately noticed that RDF was not expressive enough to create ontologies so work started to extend the language. In 1998, the W3C began working on Resource Description Framework Schema (RDFS)[9] an extension over RDF consisting of more expressive constructs such as classes, properties, ranges, domains, subclasses, etc. However, RDFS was still rather primitive and users required even more expressive power to perform automated reasoning. Because of this, two other extensions emerged around the same time; the Defense Advanced Research Projects Agency (DARPA) created the DARPA Agent Markup Language (DAML)[10] and the EU's Information Society Technologies (IST) project called OntoKnowledge [92] created the Ontology Inference Layer (OIL). Both languages served a similar purpose however DAML was based on object-oriented and frame-based knowledge representation languages whereas OIL was given a strong formal foundation based upon description logic. It soon became obvious that both efforts should be combined and a United States of America (US)/European Union (EU) joint committee[11] was subsequently setup aimed at creating one Agent Markup Language. Eventually, they created a unified language called DAML+OIL [125]. This language was further revised in 2001 by a group setup by the W3C called the "Web Ontology Working Group" and in 2004 the Web Ontology Language (OWL)[20] was created. In 2009, OWL too went through major revisions resulting in a new version of the language called OWL 2[12] which promises (amongst other things) to improve scalability and to add more powerful features. Ever since the creation of the first ontology language, different disciplines started developing their own standardised ontologies which domain experts can use to annotate and share information within their field. Today, one can find all sorts of ontologies ranging from pizzas[13] to tourism[14].

---

[9] http://www.w3.org/TR/rdf-schema/

[10] http://www.daml.org/

[11] http://www.daml.org/committee/

[12] http://www.w3.org/TR/owl2-new-features/

[13] http://owl.cs.manchester.ac.uk/browser/ontologies/653193275/

[14] http://www.bltk.ru/OWL/tourism.owl

2.3   Gluing Everything Together

However, having all the technologies and standards without having the tools that make effective use of them is useless. There have been various attempts towards defining what makes up a SW application. [139] defines a SW application as a web application which has the following features:

Semantics have to play an important role in the application, they must be represented using formal methods (such as annotations) and the application should be capable of manipulating them in order to derive new information.

Information Sources should be collected from different sources, must be composed of different data types and the data must be real (i.e, not dummy data).

Users of the application must get some additional benefit for using it.

Open world model must be assumed.

In fact, a number of prototypical systems have been designed yet they still lack a number of fundamental features. The basic and most important feature lacking in most systems is the generation of annotations automatically. Manual annotation is without doubt a burden for human users because it is a repetitive time consuming task. It is a known fact that humans are not good at repetitive tasks and tend to be error prone. The systems that support some sort of learning do so in a batch mode whereby the learning is not managed by the application but rather by the user of the system. This can be seen clearly in tools such as MnM [81], S-Cream [115] etc whereby a user is first asked to annotate and then an IE engine is trained. There is a clear distinction between the tagging phase and training phase. This has the adverse effect of interrupting the user's work since the user has to manually invoke and wait for the learner in order to learn the new annotations. Apart from this, since the learning will be performed in an incremental way, the user will not be certain whether the learner is trained on enough examples considering the sparseness of the data normally dealt with. It may also be difficult for the user to decide at which stage the system should take over the annotation process, therefore making the handing over, a trial and error process. Research towards making the annotation process semi-automatic [57] or rather fully automatic [51] [87] [47] in order to semantically annotate documents is underway and the next chapters will look into these applications.

2.4   Conclusion

This chapter explored the concepts behind the SW and clarified why it is so important. It was noticed that a large part of the technologies to make the SW possible already exists. Standards have evolved from the powerful yet difficult-to-use SGML to much more usable XML and all of its vocabularies like RDF, OWL, etc. The information needed is available in the web pages. Browsers became much more sophisticated than the original Mosaic[15] allowing customisable styles, applets, any kind of multimedia, etc. However the bottleneck seems to be related to the annotate process especially when dealing with different and diverse formats. The next chapter will deal with this issue.

---

[15] http://archive.ncsa.uiuc.edu/SDG/Software/Mosaic/NCSAMosaicHome.html

Chapter 3
Annotating Different Media

The combination of different media elements (primarily text and pictures) on the same document has been around for various centuries[1]. The idea of combining various media elements together first appeared in [43] when Bush explained his idea of the Memex. Eventually with the development of computers, most documents were text based and very few programs (apart from the professional desktop publishing systems) supported the insertion of multimedia elements. This is not surprising when one considers that the text editors available at the time could not represent layout together with the text being written. In fact users were requested to enter special commands in the text to represent different typefaces, sizes, etc. This code was eventually processed and the final document (including all the layouts, pictures, etc) was produced. In the mid-seventies, [143] created a What You See Is What You Get (WYSIWYG) text editor called Bravo. However, this was never commercialised but according to [168] a similar product based on Bravo was released with the Xerox Star. Eventually multimedia took off; word processors soon became WYSIWYG and allowed images to be inserted within documents. Web browsers brought forth a further revolution; since their target was not the printed media but the digital domain, multimedia was not limited to static content but it could also include animations, movies and sound. This obviously creates a fertile domain for new applications of annotations. The following sections expand further on these applications. According to [184], the main components of multimedia include text, graphics, images, audio and video. All of these will be covered apart from text since it will be mentioned in order sections of this document.

3.1 Different Flavours of Annotations

Annotations come in different forms or flavours, the differences are mainly dictated by the application which implements them. However, in principle we can group the different annotations in the follow categories.

---

[1] One of the oldest printed texts which includes pictures was the Diamond Sutra as described in [227].

A. Dingli: Knowledge Annotation: Making Implicit Knowledge Explicit, ISRL 16, pp. 25–32.
springerlink.com   © Springer-Verlag Berlin Heidelberg 2011

3.2 Graphics

The term graphics is used and abused in different contexts. However, for the sake of this section, we are referring to all those objects created using geometrical shapes (such as points, lines, etc) normally referred to as vector graphics. The applications of these kind of graphics range from the creation of simple pictures [126] up to the mapping of complex 3D landscapes [136]. Since vectors are so versatile, we can find various usages of annotations.

Products designed using vector graphics can be easily shared amongst different people working in a collaborative environment such as in [129] and [197]. These people can collaborate together to the creation of the product by inserting "floating" annotations attached to different parts of the 3D model under review. The strength of these annotations is that they are attached to the physical characteristics of the object rather than to a flat 2D surface.

3D modellers go through a tough time when they need some feedback from other stakeholders. Most of the time, they create a physical model of their virtual creation and circulate it around the various stake holders for comments. The result is a physical model full of scribblings. The modellers would then need to modify the 3D model, recreate another physical model and circulate it again. The cycle continues until all the stakeholders are happy with the resulting model. This happens mostly when they are creating complex models such as new buildings, aeroplanes, etc. [202] proposes an alternative to this by making use of annotations. In their approach, annotations can be attached to the virtual model, however these annotations are not simply comments but actions which modify the 3D virtual model. So what happens is that the virtual model is circulated. Using a digital pen, the different stakeholders add annotations, which when applied, can modify the virtual model. Different stakeholders are capable of seeing the annotations of others and comment on them. The final task of the modeller is to get the model with the annotations, analysis the different annotations and accept or reject them in order to produce a unified model.

Mixed reality merges together the real world with the virtual world in order to produce new enhanced views. This is the approach taken in [93] whereby the user is immersed inside the virtual world and the system allows him to interact with the virtual model using special tools (such as light pens, etc). However, these tools are not limited to just modifying the object or the view but the user can also annotate the virtual model.

Artificial Intelligence approaches too help in the creation of vector annotations such as in [220]. In this application, they make use of different techniques to annotate piping systems (such as those in waste treatment facilities, chemical plants, etc). Production rules about the different components and the relationship between them help in the labelling of the different pipes. The result of this is a hierarchy of annotations created automatically after applying inferencing on the pipe structure in operation.

Geographical Information System (GIS) are based on vector graphics too. As shown in [211], they can hold multiple layers of details (such as points of interests, road signs, etc) on the same map and these details are expressed using various different annotations.

3.3 Images

For the sake of this section, the term image refers to raster graphics. These kind of graphics are made up of a grid of pixels[2] having different colours which when combined together, form a picture as can be seen in Figure 3.1. This technology is widely spread, especially with the advent of digital cameras which are capable of creating raster images directly. The applications of these kind of graphics range from photography [130] up to the creation of 3D medical images [208]. Since raster graphics are so widely spread, we can find various usages of annotations.

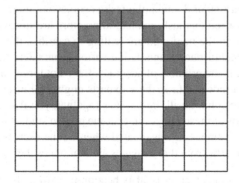

Fig. 3.1 An example of a grid of pixels used to draw a circle

In the medical domain different professionals can view medical images and add annotations to those images. In [179], a radiologist is asked to analyse some images and express his opinion about them. As soon as he notices some abnormalities, he simply adds annotations to the area under review. At a later stage, automatic annotations can also be added (as shown in [100]) and the images are queried just like a normal image retrieval engine. These systems even go a step further since these automatic image annotators might also manage to identify abnormalities in the images and tag them for further inspection by the experts.

---

[2] A picture element.

On social sites such as Facebook[3], Flickr[4], etc annotation takes various forms. The most basic form is the manual annotation where users annotate photos with information indicating the location where the photo was taken (such as geotagging as described in [148]) or annotations identifying the people present in the photo. Obviously, this brings forth various implications as discussed in [31] since even though one might be jealous about his privacy, someone else might still tag him in a photo without his consent. Another interesting aspect is the psychological one. Manual annotation is a very tedious task and in fact, a lot of projects spend incredible sums of money to employ manual annotators. However, on the social sites, annotations are inserted freely by the users. Future chapters will delve into this topic and explore why people provide free annotations on social sites even though the task at hand is still a tedious one. Other systems such as [210] try to go a step further by semi-automating the annotation process with regards to people in images. In this case, a user is involved to bootstrap the process and then a computer takes over the annotation task. Other approaches such as [48] try to eliminate the user completely from the loop by making the whole process fully automated.

Microscopic analysis is another domain where annotation is extremely important. [4] describes a framework designed to semi-automatically annotate cell characteristics. In this particular domain, the task is somewhat more complex because apart from being a tedious task, manual annotators are not found easily. They have to be experts in the field who are willing to sacrifice their time in order to go through a myriad of photos annotating various characteristics. Another issue which might arise is the problem of accuracy. Let's not forget that humans err, so after performing a repeated task, these experts might still insert some erroneous annotations. If you combine all these issues together, you'll soon realise that the manual annotations of these images in genome related studies is cost prohibitive. Because of this, a framework was created whereby the users annotate a few images, the system learns from those images and annotates the rest.

Object identification and detection is also becoming extremely useful in today's world. The idea of living in smart environments such as homes and offices is catching up, thus computers have to understand the world in which people live. Several researchers (such as [132], [233] and [193]) are working on this and trying to automate the whole process. There are various issues such as viewing a partial object or viewing the same object but from different angles. Even minor changes in the ambient lighting might influence the accuracy of the object identification process. Once these objects have been identified, the system tags them for later use, updates its own database and also infers new information based upon the facts just acquired. These facts might include relationships between objects such as the spatial relationships mentioned in [124]. These spatial relationships are derived from the pictures and allow us to learn new world knowledge such

---

[3] http://www.facebook.com
[4] http://www.flickr.com

as the fact that food is placed in a plate and not vice-versa. The potential of this approach is very promising and might lead towards changing the way we interact with computers forever since computers will be capable of understand our real world objects and how they are used.

3.4 Audio

Annotation of audio is interesting because even though it can be visualised (normally in the form of a sound wave) one cannot appreciate it until it is heard. Even when it is being played, full appreciation only occurs when the whole composition (or a substantial part of it) is played. In fact, individual sound elements have no particular meaning on their own whereas a small sequence might only give you a taste of what is to come. It is similar to seeing a picture a pixel at a time. A pixel on its own is meaning less whereas a small group of pixels might give you a small clue. However, when different pixels are combined together, they form a picture. Similarly, various sound elements combined together form a musical composition, a speech or anything audible. The major difference between the visual form and the audible form is that whereas a picture can be enjoyed by the human brain in a fraction of a second, the brain would probably take seconds, minutes or even hours to appreciate a sound composition. In our world, sounds are very important, and their application range from being the main communication channel used by humans up to the unthinkable Acoustical Oceanographic research as specified in [158]. In the following subsections, we'll have a look at how sounds have been annotated and why.

In music [120] and [17] mention various possible annotations. First of all, the annotations have to be divided into three; those within the file, those across different files and those shared amongst different people. The first kind of annotations too can be further subdivided. Some music experts might be interested in the acoustic content including the rhythm, the tonality, description of the various instruments used and other related information (such as the lyrics). To these annotations, one can also add social tags such as those mentioned in [83] which include comments about parts of the songs or even emotions brought forth by the piece of music. Annotations across different files also gather together different meta properties shared by multiple files such as the author, the genre and the year. Finally, the sharing of annotations amongst different people allows users to search for music using semantic descriptions. Through these searches, profiles can be constructed which suggest to the users which kind of music might be of interest to them. The social network will also help them identify pieces of music which are maybe unknown to them or which would not feature as a result of their normal search. Obviously, this brings about new powerful ways of accessing musical pieces.

Speech is a form of audio whereby words (rather than music) are predominant. When the audio is a monologue, the speech is converted to text (using speech recognition software) and annotated using normal text annotation methodologies. However, extra care must be taken as mentioned in [85] because speech normally has some subtle differences from written text. In fact, it is very normal to find unfinished sentences, ungrammatical sentences and conversational fillers[5]. When two or more people are speaking, it is normally referred to as a dialogue. In this case, the situation is somewhat more complex because apart from the issues mentioned so far, one has to add others such as ellipsis, deixis and indirect meanings such as ironic sentences.

Semantic taggers such as the ones described in [66] and [36] are also used to annotate speech (once it is converted to text). These taggers identify semantic information such as named entities, currencies, etc. The interesting thing is that they can also be easily expanded by using gazetteers and grammars [67]. The semantic information is generally associated with an ontology which gives the information a grounding relative to a world domain thus avoiding any ambiguities. When it comes to dialogues, the same techniques used in annotating speech are adopted. However, in this case, we can also annotate dialogue acts. These dialogue acts taggers such as the ones described in [195], [159] and [154] are capable of identifying speech acts within a dialogue. These speech acts label sentences or phrases as being a question, a statement, a request or other forms. The amount of dialogue acts used can vary according to the system being used [206].

3.5 Video

The term video generally refers to the transmission of a series of pictures displayed one after the other in quick succession (which gives the impression of movement) combined with synchronised sound. Obviously this might sound as being a restrictive definition of video however it encompasses the basic principles of the technology. Today's technologies have made giant leaps in quality when it comes to sound and images. Companies such as Dolby[6] and THX[7] provide the audience with an impressive experience. Most of these systems make use of multiple speakers to play sounds from different directions. Images too have reached High Definition and are now slowly venturing into the 3 Dimensional domain [230]. With the advent of camera phones, the creation of video has been widely adopted and today, the use of this technology ranges from the creation of simple home made videos up to the impressive Hollywood blockbusters.

When it comes to annotation, video has been quite held back mainly because of the lack of automated methods to tag images. Several projects have been trying to annotate videos such as:

---

[5] Phrases like **a-ha**, **yes**, **hmm** or **eh** are often used in order to fill the pauses in the conversation. It normally indicates periods of attention or reflection.

[6] http://www.dolby.com

[7] http://www.thx.com

Detection and tracking of objects in [63] whereby the researchers are managing to identify predefined objects and track their movement throughout the video. This can be used to identify moving targets such as cars, annotate it and track its movement across a landscape. Similar techniques are used in other domains such as for video surveillance [224], for animal identification [189], etc.

Annotating sport events from video such as the work conducted by [218] and [22] whereby events in a football match are automatically annotated by the system and recorded. It is interesting to note that apart from handling different errors (brought forth by the quality of video) the system must also consider the rules of the games and ensure that the annotations adhere to those rules. Similar researchers studied these techniques and applied them to other sporting events such as tennis [237]. The benefits of these annotations is unimaginable since they are capable of creating a transcript of the match almost in real time thus enabling people unable to watch (such as while driving or people suffering from some form of disability) to understand what's happening.

Movies too need annotations. The most common form of annotation found in all DVDs are the subtitles. These are essentially a transcript of the dialogue displayed in synchronisation with the movie when it is being played. [95] and [21] created techniques which lists cast members in a movie and [89] goes a step further by identifying the names of the characters appearing in the movie and annotating when they appear. Obviously, understanding what's happening in the movie is somewhat more complex. When you consider a sporting event such as those mentioned before, the rules are fixed so the actions are predictable and finite. In a movie, there's no predictable plot and unexpected twists improve the movie. This makes the task of creating annotations for movies harder. Notwithstanding this, there has been various attempts to identify actions such as [144] which tries to categorise movie scenes involving people walking, jogging, running, boxing, waving, answering the phone, hugging, sitting, etc. However, these systems are still really far from understand what is actually going on.

With the rise of Web 2.0 technologies and the proliferation of online videos thanks to sites such as YouTube[8] and projects such as Joost[9], annotations in videos are gaining a more prominent role.

YouTube allows users to enter different annotations on top of the videos. These range from adding additional information to particular scenes or just comments. These annotations can also include links thus allowing users to branch from one scene to another. This is interesting because it disrupts the linear structure of movies using hypermedia whereby the links can lead to all sort of media elements. YouTube provides four types of annotations; those in speech bubbles, those as notes, spotlights which highlight areas of a movie (which only reveal the text when the mouse moves over them) and video pauses which pause the movie for a specified period of

---

[8] http://www.youtube.com
[9] http://www.joost.com

time in order to reveal the annotations. Joost on the other hand is both a desktop and a web application whose aim is to create a new form of TV which is full of multimedia elements (and not just movies), which is on demand and interactive. Joost too allows the insertion of annotations in the movie by making use of the social network. In fact these annotations can be shared between different groups of people and the annotation can also be obtained from other online sites thus integrating information from multiple independent sources together. The annotations in Joost are not simply limited to text as in YouTube but they can also include freehand scribbles.

3.6   Open Issues with Multimedia Annotations

This chapter has shown that annotations are extremely important irrespective of the media being used. However, even though various solutions exists, there are still several open issues which need to be dealt with.

- A lot of media tools still do not support annotations [175]. In fact most of the annotations are added by third party applications and are stored outside the media file rather than being integrated within. This obviously has its pros and cons however, a tighter integration would definitely be beneficial.

- Even if the tools catered for links, the link between the media data and the annotations is not so straight forward. An annotation can refer to a whole media document, to a subset or even to a single element within that document.

- The lack of standardised annotation vocabularies makes annotations hard to reuse. If someone would go through the hassle of developing such vocabularies, it would take a lot of time and cost huge sums of money. In the end, there's no guarantee that these vocabularies would be adopted. There have been various attempts at achieving this such as [99] however so far, no consensus has been achieved.

- The uncertainty as described in [37], introduced by automated annotation processes is something which can deplete the value of the multimedia document rather than enrich it. As an example, if a document about sports is wrongly annotated with information about finance, its relevance will be severely impacted and it would be hard for it to feature in relevant searches.

3.7   Conclusion

This chapter explored the various annotation systems which handle different multimedia elements. It is immediately evident the value of these annotations and how they can be used to enrich the user's experience. Not withstanding this, there are still quite a number of open issues which need to be addressed. The next chapter, will look at the actual annotation process, in particular how manual annotation is being performed.

Part II
Leaving a Mark ...

"The mark of a good action is that it appears inevitable in retrospect."

Robert Louis Stevenson

Chapter 4
Manual Annotation

The task of annotating content has been around for quite a while and this is clearly evident from Chapter 1. Throughout the years, various tools were created which allowed users to annotate documents manually. This chapter, will survey the various tools available, will divulge into their uses, their potential but also their limitations. Then it will explore the various issues associated with manual annotation.

4.1 The Tools

In itself, manual annotation does not require any particular tool when dealing with physical documents. However, the situation changes when we handle digital documents because without additional support, annotation is not possible. We have already seen how digital annotations started with the development of SGML and the creation of the LaTeX system. However, wider adoption of annotations was meant to come with the Xanadu project, but we missed the bus!

Xanadu[169][170] was an idea of Professor Ted Nelson, the person who is accredited with coining the term hypertext[1]. Xanadu represents his idea of the WWW long before the current web was even conceived. Fundamental to this web was the idea of the hyperlinks where every document can contain any number of links to any other document. Essentially, this made it possible for the first annotations to appear since document content could be linked to other content by using hyperlinks. The major difference between these hyperlinks and what we have today is that the hyperlinks are not stored in the document itself for two reasons. First of all, different media types would make the embedding of annotations within the document difficult. Secondly, a document can have an infinite number of annotations. In extreme cases, the annotations would outweigh the actual document and obscure its content. For

---

[1] This term was first used in a talk which Professor Nelson gave in 1965 entitled **Computers, Creativity, and the Nature of the Written Word**. A news item of the event can be found at http://faculty.vassar.edu/mijoyce/MiscNews_Feb65.html

A. Dingli: Knowledge Annotation: Making Implicit Knowledge Explicit, ISRL 16, pp. 35–42.
springerlink.com   © Springer-Verlag Berlin Heidelberg 2011

this purpose a system called Hyper-G[19] was developed but in reality, it was never popularised.

NoteCard[112] is one of the first hypermedia systems available. It was created at a time when the WWW was still being conceived and hypermedia was just a small topic segregated between different university campuses. The idea behind NoteCard was to create a semantic network where the nodes are electronic note cards and the links are typed links. The system allowed users to see the links, manipulate them and navigate through the network. An electronic card had a title and could contain information of various forms such as text, drawings or bitmaps. Other types of note cards could be constructed from the combination of the basic types. The system also had a browser, a program rather different than the web browsers we have today. Its task was to visualise the network of note cards. When several note cards were linked together, they were collected in a file box which is equivalent to a modern day folder. Even though note cards were used to annotate other documents, essentially, they were the precursor of today's WWW.

ComMentor[190] was one of the initial architectures designed to handle annotations on web pages. At the time, the most popular browser was the Mosaic browser so ComMentor had to allow users to insert annotations in the browser. The annotations were divided into three groups, private (visible only to the owner), group (restricted to a group of people) or public (available to anyone). New annotations could be inserted and viewed, however the system did not cater for edits. The system also separated the annotations from the content thus ensuring that the original document is not modified in any way. At the time, the rational why users needed annotations was very similar to what users need today. However there were two additional reasons worth mentioning, according to the creators of the system, annotations could be used to track document usage. Thus if a particular group of people did not manage to view a rather important document, they can be notified about it via email. The second reason is to give the document a Seal Of APproval (SOAP)[69]. The seal is a rating system used to describe the importance and validity of a document.

CoNote[71] was a system created back in 1994 aimed at supporting cooperative work system. The idea was to allow a group of people working together to annotate document and share the annotations between them. Such a system was tested in a classroom environment whereby students and teachers could share their comments, notes, etc. With difference from generic annotation systems, this system was based around a context (which was the document) and people commented around that fixed context. In actual fact, the original document is not modified since the system stores the annotations remotely on a server. This has the added benefit that every document can be annotated (even those that are read only). The positioning of the annotations is also restricted to specific points which can be chosen by the author of the document or by an administrator. When an element of the document is annotated several times, the annotations are showed as a thread thus allowing for easy viewing. Users can also search through the annotations inserted by using attributes such as the date when the annotation was inserted, the authors, etc. It is interesting to note that experimental results have shown that the educational experience provided to the students was greatly enhanced by the system. Students were seen annotating documents and commenting or replying to annotations created by other students. This was one of the first attempts at creating what is today known as the social web.

JotBot[219] is a prototypical system with the scope of annotating web pages, however, most of the work is performed on the client side rather than the server. This is achieved by making use of a Java applet whose task is to retrieve annotations from specialised servers and presenting a comprehensive interface to the user. Since the annotations happen on the client side after the page is downloaded, this can be considered as being one of the first on-the-fly annotation tools. An interesting concept used in JotBot is that annotations are not associated with a document for an indefinitely amount of time. In fact, they all have an expiry date and users can vote to extend the life of worthy annotations. This creates a sort of survival of the fittest approach whereby only the most relevant annotations (according to the users) are kept.

ThirdVoice[153] was a commercial application launched in 1999. The idea was to create a browser plug-in capable of annotating any page on the internet. The original content was never altered, in fact, the annotations were inserted after the web page was rendered by the browser. However, as soon as the service was launched, it was immediately unpopular with a lot of web site owners [142] and some of them even defined it as web graffiti. A lot of these people were afraid of the idea of having people distribute critical, off topic and obscene material on top of their site. Some of the web site owners even threatened the company with legal action however in reality, no one ever filed a law suite. Another issue arouse from the annotator's side since the annotations were stored on a central server controlled by Third Voice thus causing a potential privacy issue. Ironically, the company's downfall was not due to these issue but rather to the dot-com bubble [180]. At the time when the owners were going through another round of financing, the internet bubble was bursting so investors were weary to invest in internet companies.

Annotea[135] is an web based annotation framework based on the RDF. Annotations are considered as being comments inserted by a user on a particular website. These annotations are not embedded within the document but are stored in an annotation server thus making them easily shareable across different people. These annotation servers store the annotations as RDF tripples thus essentially they do not use normal databases but tripple stores [231]. Apart from storing annotations remotely, Annotea also allows the storage of annotations locally in a separate HTML file. The annotations make use of different standards which include a combination of RDF, the Dublin Core[2], XML Pointer Language (XPointer)[3] and XLink[4]. The

---

[2] http://dublincore.org

[3] http://www.w3.org/TR/xptr

[4] http://www.w3.org/TR/xlink

framework itself proved to be quite popular and in fact, it is implemented by several systems amongst which Amaya[5], Bookmarklets[6] and Annozilla[7].

CritLink[134] was a proxy based annotation system. Users could access an initial page known as the Mediator (which was originally located at http://crit.org) and request a specific location. The job of the Mediator was to retrieve the page, annotate it using the annotations stored in its database and present the modified page to the user who posted the original request. The system used a series of pop-up windows to display both the existent annotations and the control panel through which new annotations could be added. Unfortunately, the system did not last long due to two particular reasons. First and foremost, the back end suffered from a series of hardware failures. This shows the risks which centralised annotation servers pose whereby a single failure can effect the whole system. It also provides no redundancy to the users thus if the harddisk fails, all annotations stored on the disk are lost. The second problem was related to abusive annotations. Since users are capable of annotating any page, this leaves scope for abuses.

The Annotation Engine[8] is similar in principle to other tools mentioned in this section however, it has some subtle differences. The tool was originally inspired by CritLink and it works as a proxy. All URL requests are sent through the proxy but when they are retrieved, they are modified by inserting the annotations and only the modified version of the document is displayed. With difference to other methods, the annotations are physically inserted in the document before it is being sent to the user thus making them an integral part of the document rather than merely a layer on top of it. This makes the annotation engine a rewriting proxy. The annotations inserted are similar to footnotes referenced by a number and referencing a link. However, when the users click on these links, the details of the annotations are displayed in another frame. The advantage of this is that the system is rather fast since the manipulation occurs on the server and the annotations can be applied to virtually any HTML document. However, the downfall of this approach is that since the original HTML is being modified, the program can have some undesirable effects on the design of the page. This can be the case with Cascading Style Sheets (CSS) where the layout is separate from the content, thus the proxy is not aware of the CSS and the colour of the annotations can easily clash with the colours used on the page. In addition, the use of frames is not desirable since frames can cause several problems[9] related to bookmarking, searching, navigation, coding, etc.

MADCOW[34][35] is an annotation system implemented as a toolbar on the client's side coupled with a server holding the various annotations. When a page is accessed by the user, the toolbar annotates the page based upon the annotations

---

[5] http://www.w3.org/Amaya

[6] http://www.w3.org/2001/Annotea/Bookmarklet

[7] http://annozilla.mozdev.org

[8] http://cyber.law.harvard.edu/cite/annotate.cgi

[9] http://www.yourhtmlsource.com/frames/goodorbad.html

stored on the server. Tooltips together with pop-ups are used to display the various annotations. The interesting thing about this system is that they claim to offer multimedia annotations. In fact, multimedia elements such as pictures can be annotated too. Apart from this, they can also be used as annotations themselves. In an interesting case study present in [35], the authors showed how MADCOW can be used as a collaborative tool by art restorers. Different users contributed various comments to different parts of the document. However, an issue arouse on a picture of a particular room and one of the restorers annotated the picture with another picture showing an artistic impression of how he imagined the refurbished room. Obviously, annotations are not only bound to the original document but also to other annotations. In fact other users were commenting about the artistic impression of the new room. MADCOW too provides several privacy options and allows users to search through the database of annotations.

WebAnn[30] is a shared annotation system which caters for fine-grained annotations. The original context was the class room whereby users could share educational content and add comments about different aspects of the document. Since comments were not added to the whole document but to parts of it, this anchored the annotations to specific elements thus placing annotations within a well defined context. The class environment required a system of annotations which could be shared and also allowed for threaded discussions. The system displayed annotations alongside the text in separate frames. The threaded system allowed questions to be asked and answered, it allowed the identification of issues, the writing of opinions and the handling of discussions. The studies performed on WebAnn showed some interesting results, first of all, it seems that students generally prefer to use a newsgroup rather than an annotation system to discuss these matters. However, it transpired that those that actually used the system where much more productive than their counterparts, in fact, their contribution to the topic was twice as many comments as one would expect in a newsgroup. It seems that comments grounded directly to a context helps a community create richer discussions.

Collate[215][15][98] is an acronym for Collaboratory for Annotation Indexing and Retrieval of Digitised Historic Archive Material. Similarly to WebAnn, Collate is a research tool created for a particular community of users, in fact it is aimed at helping researchers in the humanities. Users interested in historical film documentation dating back to the 20's and 30's can collaborate together to create annotations to censorship documents, press material, photos, movie fragments and related posters. The system makes use of well defined typed links which can occur either between the document and the annotations or in-between the various annotations. What's rather interesting in this system is the way annotations are treated. In fact, annotation threads whereby different annotations are inserted to explain other annotations are considered as being a part of the document and not just external links. The idea is that these kind of annotations create a discourse context which is interlinked to portion of the text. Irrespective if the arguments brought forth are coherent or not, the fact that different experts are debating different theories associated with that portion of the document is enough to enrich the original document since such information can provide other users with additional viewpoints of the same document. Because of this, it is considered by the creators of the system as an integral part of the document. This information together with the type of annotations and the position within the document is later used by the users to search throughout the collection of documents.

FAST - Flexible Annotation Service Tool[5][6][8][9][7] is an architecture designed to support different paradigms such as Web Services, P2P, etc combined with a Digital Library Management System. Similarly to other annotation systems, FAST support both user and group annotations. In fact, every annotation can be either private, public or shared. FAST was designed to be rather flexible thus freeing it from any particular architectural constraints. This flexibility creates a uniform annotation interface irrespective of the underlying databases. In so doing, a switch between different architectures becomes transparent to the user. The importance of this is that the annotations can be easily stored in different databases simultaneously. This brings us to the idea that a document might posses an infinite number of annotations which would be impossible to visualise. As an example, a web page about rabbits might be annotated with information about pets, discussions by vets and instructions on how to prepare rabbit recipes. Obviously, different people might be interested in only a small subset of those annotations. So a cook accessing the page would only be interested in the recipe related annotations. These different dimensions on the same page, brings about the need of categorising annotations and show or hide them when appropriate. However there might also be cases when the different dimensions need to merge. A user having a pet rabbit might need to check and eventually link to the vet's annotations related to the well being of the animal.

4.2   Issues with Manual Annotations

As we've seen in this section, users are practically spoilt for choice when it comes to manual annotation. Not withstanding this, manual annotation suffers from its own set of problems.

First of all, annotating documents manually is costly and eventually time-consuming. Humans have two major flaws when it comes to annotations. First and foremost, they have a very limited attention span. [65] claims that the maximum attention span of an adult is about 20 minutes. When this time elapses, it can be renewed if the person is enjoying the experience. Not withstanding this, the more it is renewed, the less effective it becomes (unless the person takes a break). This mean that when a user annotates a document, since the attention span is rather limited, the task at hand will become relatively harder with time. The second flaw is that humans commit errors. People are different than software agents because they are not capable of repeating the same process precisely as machines do. These errors are further accentuated when the attention span declines. The combination of these two factors makes the whole process timely and eventually costly. Even though various techniques have been developed to facilitate the annotation process, most applications require higher level annotations which are only possible by using human labour. Finally, if the domain being annotated is highly specialised (such as the annotation of the legal documents), there would be very few people who can understand the documents and annotate them, thus increasing the annotation costs even further.

Secondly, human annotation is highly subjective. Different domains have different experts and sometimes, these experts do not agree on the underlying theories [196]. Even if they do agree, different people tend to interpret document differently and in so doing, creating inconsistencies within the same document collection. The best approach to solve this issue is to have several people annotating the same document and use those annotations to calculate an inter-annotation ratio in order to evaluate the validity of the annotations. However, this is not always possible due to various constraints (time, costs, etc). Time is another important factor which plays upon the subjectivity aspect. Back in the nineties, astronomical annotations related to our solar system would have marked Pluto as being the ninth and most distant planet in our system. However, a few years ago, the scientific community changed the definition of a planet (as per [203]) and in the process, demoted Pluto to a dwarf planet. This clearly shows that correct annotations might not hold the test of time and their validity might need to be reevaluated. In this example, there was a change in definition brought forth by a scientific community, however, changes can be much more trivial. A person annotating a document might do so by considering particular viewpoints. Some time later, the same person or even someone else might require the same document but with radically different annotations.

This brings us to the third issue, restrictiveness. Annotations can be a little bit restrictive if we use formal metadata which can be easily understood by machines. This is why annotation tools are important because they provide users with a high level of abstraction thus hiding away any complex formalism. On the other hand, if free text is used because it is much more natural for humans, we are faced with the opposite problem because it would be very hard for machines to interpret those annotations. That is why the Semantic Web and its technologies are extremely important because according to [29] and [228], annotations should be both machine and human readable thus solving the problem once and for all.

The forth issue has to deal with rights and privacy issue. A person annotating a document might be considered as someone adding intellectual content to the document. Thus, since he is enriching the document, some issues might arise about who owns the rights to those annotations. The other issue is related to privacy. Some data in the document might include private or sensitive information which must be handled with great care. This is very common with medical records whereby the personal details are stored together with the medical history of the patient. Even though annotations would be very useful especially to discover interesting correlations between personal data and the medical history, the fact that humans annotated these records exposes them to various risks.

Chapter 5
Annotation Using Human Computation

In the late 18th century, the Holy Roman Empress Maria Theresa (1717-1780) was highly impressed with a chess-playing machine known as the Mechanical Turk [194]. This machine was created by Wofgang von Kempelen and it possessed a mechanism capable of playing a game of chess against a human opponent. In reality, this was nothing more than an elaborate hoax [229] having a human chess master hiding inside the machine. This illusion lasted for more than 80 years and it baffled the minds of distinguished personalities such as Benjamin Franklin and Napoleon Bonaparte. The point behind this story is that at the time, machines were not capable of playing a chess game and the only way to do so was to have a person acting as if he was the machine. This is once again accentuated in the novel, the Wonderful Wizard of Oz [25] whereby the wizard is nothing more than a mere mortal hiding behind a curtain and pretending to be something much more powerful than he actually was. The same approach is also normally used in annotation tasks as well. When a machine is not capable of annotating a set of documents (E.g. images), the task can be outsourced to a human in order to solve the annotation problem, this is generally referred to as human computation.

Modern human computation was first introduced in a program found on the CD attached to [72]. In this program, the user can run a genetic algorithm[1] and the user acts as the fitness function[2] of that algorithm. In recent years, Amazon.com too took over a similar initiative and in fact they also named it the Amazon Mechanical Turk[3]. The idea was to create a marketplace whereby tasks, which are difficult to perform using intelligent agents, are advertised on this site and users willing to perform the task can propose to perform it. However, all of these approaches are not enough when we are faced with huge tasks such as annotating large volumes of documents. The following sections, will look at how the network is helping us solving these tasks by using shared human computation.

---

[1] A genetic algorithm as defined in [164] tries to solve optimisation problems by using an evolutionary approach.

[2] A fitness function is an important part of a genetic algorithm which measures the optimality level of a solution.

[3] https://www.mturk.com/

A. Dingli: Knowledge Annotation: Making Implicit Knowledge Explicit, ISRL 16, pp. 43–58.
springerlink.com

5.1   CAPTCHA

Imagine you were assigned the task of annotating a large collection of documents. If a machine was intelligent enough to perform the task, it would start annotating immediately without any complaints and irrespective of whether the task is overwhelming. Unfortunately, since we do not have machines with such intelligence we have to rely on humans. But a human faced with an overwhelming task will probably give it a try but then walk away after realising that it is something impossible to achieve on his own. Since computers are good at some things whilst humans at others, the idea is to combine these two strengths together in order to achieve a greater goal. A system that can be used to achieve this is the Completely Automated Public Turing test to tell Computers and Humans Apart (CAPTCHA)[10][11] a system designed to perform a test which distinguishes between automated bots and humans. Although at first sight it might look like a test unrelated to annotation, what we're interested at is the side effect of this system.

A CAPTCHA is generally a picture showing some distorted text. Current programs are not capable of understanding what's written in the text but a human can easily do so. So to pass the test, a user simply types in the textual equivalent of the text in the picture. When Google created a CAPTHCA system which it called re-CAPTCHA [221], it decided to make use of the text not just for testing purpose but also to generate annotations.

Back in 2005, Google announced that it would embark on a massive digitisation program whereby it will digitise and make available huge libraries of books[4]. Obviously, controversy broke out about rights issues however this was partially sorted through various deals with writers, publishers, etc. Initially, prominent names such as Harvard, Standford, Oxford and many others took the plunge. Even though huge sums were invested in this digitising project and new devices were created capable of turning pages automatically without damaging the original document, the bottle neck was the error rate of the Optical Character Recogniser (OCR). Irrespective of the various improvements in OCR technologies, there are still parts of the document which cannot be translated to text automatically. This might happen for various reasons such as the document might be old, damaged, scribbled, etc. This is where re-CAPTCHA comes into play. Through the digitisation process, Google engineers can generate two lists of words, one containing words which were recognised successfully and the other containing words which were unknown to the OCR (essentially those where the error rate is very high). In order to generate the CAPTCHA, they gather an image of a word from every list, distort it and display it to the user. The user is then asked to identify both words. Based upon the user's answer, a rating is given to the unknown word. So if the user manages to recognise the known word and write the textural equivalent, his answer for the unknown word is taken as being correct as well. The same idea holds for the inverse, if the user misspells the word known by the system then the unknown word is also considered as being wrong. Obviously such an approach is not fool proof. However, the experiments documented in [221] show some impressive results. In fact, they claim that their system which

---

[4] http://books.google.com/googlebooks/library.html

essentially makes use of humans to annotate distorted images of texts manages to solve about 200 million words a day having an accuracy of 99%. Essentially, this is nothing more than manual annotation used effectively, in fact, everyone who manages a website can place reCAPTCHA on their website by following simple examples. If we were to ask how Google manages to solve all those words each day, the answer is two fold. First of all, the reCAPTCHA system serves a very useful purpose so websites use it. Secondly, the task takes only a small amount of time from each user so users don't mind using it. However, there's another reason apart from usefulness, that would entice people to annotate content and this is to entertainment themselves.

5.2 Entertaining Annotations

Gaming is one of the biggest markets available online. Millions of users play online games, in fact, according to [157], people spend a total of 3 billion hours per week playing online games. She also suggests that since gamers spend so much time immersed in serious gaming, their efforts should be placed to better usage. This is in essence what the following systems do. They provide an entertaining and competing environment whilst creating annotations as a byproduct of the system.

5.2.1 ESP

[12] was one of the first games designed with the purpose of annotating images. The idea is rather simple and similar to the CAPTCHA. Several users log into the system and decide to play the ESP game. Two random users (unknown to each other) are paired together and they are presented with the same image. Their task is to provide a common label for the image within a specific time frame. If they manage to propose the same label, the system gives them some points and shows them a different image. This process continues until the timer ends. Essentially, their task is to guess the label which the other user might insert. This is why the game is called ESP (short for Extra Sensory Perception) since it involves a process of receiving information from someone else without using the recognised senses (there's no chat or conversation possible in the game). To make things slightly more difficult, the designers of the game also introduced some taboo words. Initially, images have no taboo words associated with them but when users start agreeing on common labels these labels are inserted in the taboo list and they cannot be suggested by other users. If the number of taboo words exceeds a particular threshold, the image is removed from the database of possible images since most probably, users can't think of other labels thus making the game frustrating. From the evaluation of the ESP system, it transpired that the game was rather fun and its 13,600 users managed to provide more than 1.3 million labels. A manual check on a sample of the labels found that out of these labels, 85% were useful to describe the image. Similarly to ESP, [146][145] launched TagATune, a game aimed at annotating music. The

usefulness of such techniques is quite evident and in fact in 2006, Google launched the ESP game on its own site called the Google Image Labeler[5].

5.2.1.1 Googe Image Labeler

The task of searching and retrieving images might seem trivial but for a search engine, it is extremely complex. A lot of research has been undertaken on the matter (see [149], Google[6], Yahoo[7],Bing[8]). Most of the approaches adopted make use of text or hyperlinks found on web pages within the proximity of the image. Another approach proposed by WebSeeker[9] combines text based indexing with computer vision, however the improvement of such an approach does not seem to be significant. The problem with all of these approaches seem to stem from the fact that they rely too much on text to determine the image tags. Text can be scarce, misleading and hard to process thus resulting in inappropriate results. This is why Google adopted the Google Image Labeler in order to improve the labels associated to the images in its databases. Essentially, the underlying approach is very similar to the original ESP however it has some subtle differences. For example, the game awards more points to labels that are specific. So an image of the Pope labelled as "Benedict" would obtain more points than the generic label "man". The game also filters abusive words, most of which are not even real words yet they were used by users to sabotage the system. However, not withstanding these and other issues, the Google Image Labeler is working effectively to help Google improve its image search thus providing users with better results.

5.2.2 Peekaboom

[13] is a game similar in spirit to ESP whereby two users are playing an online game and indirectly, annotating images. As the name suggests, one of the two users is referred to as Peek and the other as Boom. The role of Boom is to reveal parts of an image in order to help Peek guess the word. So if the system displays an image to Boom containing both a car and a motorcycle, and Peek has to guess that it is a car, Booms' role is only to reveal the car and keep the motorcycle hidden. What's happening is quite obvious, the game is not simple a remake of ESP but rather a sophistication over it. Whereas in ESP, labels are associated with the whole picture, in Peekaboom, labels are associated to specific areas in the picture thus indirectly annotating that area. The game also allows for hints and pings. Hints allow Boom to send flashcards to Peek and in so doing, help him understand whether he is after a noun, verb or something else. Pings on the other hand are a sort of signal (displayed as circular ripples which disappear with time) sent by Boom to help Peek focus

---

[5] http://images.google.com/imagelabeler

[6] http://images.google.com

[7] http://images.search.yahoo.com

[8] http://www.bing.com/images

[9] http://persia.ee.columbia.edu:8008

on specific aspects of the picture. Through this game, the system collects different kinds of data which include:

The relationship between the word and the image (i.e. if the word is a verb, noun, etc) through the use of hints.

The area of the image(including any context) necessary for a person to guess the word.

The area within the object by noting down the pings.

The most important parts of an object which is identified by recording the sequence of revelations. For example if we have a picture of President Barack Obama, revealing the face would give a good indication of who the person is whereas showing just his feet is useless to identify the person.

Poor image-word pairs which are filtered out throughout the game since their popularity will rapidly decline.

From the evaluation, two things transpired. First of all, users seem to find it enjoyable, in fact some of these users play the game repeatedly for long stretches. Secondly, the annotations generated through this system were very accurate, because of this, they can be easily used for other applications.

5.2.3 KisKisBan

Another game similar in spirit to ESP is [123] however it proposes a further refinement. Rather than having just two persons trying to guess similar tags for an image, KisKisBan introduced a third person in the game normally referred to as the blocker. His role is precisely to block, the two players collaborating together, from finding a match. This is achieved by suggesting words before they do. By doing so, those words are placed in a blocked list and they cannot be used by the players. This mechanism ensures that no cheating occurs (such as agreeing on the labels through some third party chat) between the two players collaborating together. However, the major advantage of such a system is that in every round, several labels are generated per image (and not just one as in ESP) thus making the system effective with a precision reaching the 79% mark.

5.2.4 PicChanster

[49] is an image annotation game which has two major differences from what we've seen so far, it exploits the social networking sites and it is based on a system similar to reCAPTCHA. Rather than being just a game in an applet or in a browser, PicChanster is integrated in Facebook, one of the most popular social networking sites on the internet which boasts more than 500 million active users[10]. Placing the game in

---

[10] http://www.facebook.com/press/info.php?statistics

such a context makes it easier to distribute (by using Facebook invites, feeds, etc) and use. The second major difference is the process adopted. With difference to the games we've seen so far, PicChanster is a single player game. The competing aspect of the game is derived from the social context of the Facebook sites where scores get posted to the user's profile and different users boast with their online friends about their achievements. Being a single player game, the system is slightly more complex since the user is not checking the validity of the answer with another user however, a work around was found as follows:

- PicChanster has two databases full of images and their corresponding labels, one is called uncertain and the other is called the certain. The images and the corresponding labels in the certain database were collected from sites containing manually annotated images such as Flickr[11]. Since the labels in Flickr were inserted manually, we assume that they are correct. The images in the uncertain database were harvested from popular image search databases such as Google Images[12]. These annotated images are classified as uncertain because they were collected using traditional image indexing techniques (which use the text in the document, etc) whose accuracy is rather low.

- Each game lasts for two minutes and the scope of the game is to go through a series of apparently random images and insert up to four labels per image.

- Scores are only awarded to matching labels in the certain set but the user is not aware which image comes from which set. In reality, half of the images belong to the certain set and the other half from the uncertain set.

- By using the labels retrieved from the certain set, the accuracy of the user can be rated and assigned to the labels given in the uncertain set.

- An image is labelled several times by different users and each time, the accuracy of the labelling is stored and augmented to previous ratings.

- When the image has been annotated several times (determined through experimentation) and the accuracy is above a certain threshold (which was found empirically), the annotation is shifted from the uncertain set to the certain set.

In essence, PicChaster presents a new way of annotating images without necessary requiring two or more people competing or collaborating with each other. Similarly to reCAPTCHA, not all images have been manually annotated thus providing new annotations as a side effect of the game.

5.2.5 GWAP

The creators of ESP, Peekaboom and reCAPTCHA eventually got together and created Games With A Purpose (GWAP)[13]. The idea is to have a site which collects

---

[11] http://www.flickr.com

[12] http://images.google.com

[13] http://www.gwap.com

different games whose scope is to generate different types of annotations. In fact, the site hosts the following games:

The ESP Game is a modern version of the original game described earlier.

The Tag a Tune Game is similar to the ESP game but based around tagging tunes rather than images.

Verbosity is a game made up of a describer and a guesser. The role of the describer is to help the guesser guess a secret word by giving clues. The side effect of this game is to get various descriptions for particular words.

In Squigl two people are presented with a word describing an object and an image. The scope of the game is to trace the object in the image. The side effect of this game is to associate words with objects inside an image.

In Matchin two people are presented with two images and they have to select the image they like best. The side effect of the game is to register the tastes of the person.

In FlipIt a user is asked to turn tiles and match pairs of similar images.

The PopVideo game is similar to the ESP and TagATune game but its aim is to tag videos.

5.3 Social Annotations

In the past decade, a class of websites normally referred to as social networking sites emerged and quickly gained popularity. In fact, these sites are normally found listed at the top of the list[14] containing the most accessed websites worldwide. Facebook is in second place, YouTube is third, Twitter is tenth and the list goes on.

These sites generally share some common features such as the need to create a personal profile, the facility to upload digital media, the facility to blog or microblog, etc. Amongst these features we also find social tagging. This tagging allows users to tag an item or a group of items by assigning keywords to them. These items are normally web resources such as online texts or images and as soon as they're annotated, the annotations become immediately available for anyone to use and see. Social annotations differ from traditional annotations since the tags are not based upon an ontology or a controlled vocabulary but they are freely chosen by the users. Given enough tagging, folksonomies will emerge which can easily augment or replace ontologies [200]. Because of this, [222] claims that the level of interest in manual tagging witnessed a renewal in recent years. This can be seen in the following websites where annotation, is an integral part of their business process.

---

[14] http://www.alexa.com/topsites

5.3.1 Digg

The social news website Digg[15] is one of the most visited sites online. The idea is to create a news website whose editors are the users. Essentially all they have to do is to find a story online, post it to Digg and mark it. The annotation used is a sort of vote and the more people like it, the more it rises in popularity in comparison with other news items. Each link can also be commented by using a micro-blog and these comments can also be voted just as the articles. The annotations in digg are stored in a central server thus allowing sharing between the various users. The popularity of every link is something temporary and not permanent. This is because digg simulates a dynamic marketplace where news items are constantly gaining popularity and surpassing others. Thus, because of this dynamicity of the diggs, it is highly unlikely that an article will stay at the top for a long period of time. Let's not forget that the model of a dynamic newspaper must ensure that popular news items get promoted to the top immediately. Finally, even though digg requires registration, this is only needed to customise the digg interface but not to exchange personal information with other users of the site as in other social sites.

5.3.2 Delicious

The social bookmarking site Delicious[16] is designed to store and share bookmarks. The idea is to create an online repository from where a user can access his own bookmarks irrespective of his physical location and irrespective of the device he is using to access them. This solves the problem of having a set of bookmarks locked in one specific browser on some device. The power of delicious is twofold, first and foremost the annotational aspect of the system and secondly the social aspect.

Every link, apart from the title and a description can also have tags associated to it. These tags are used to annotate the link by providing associated keywords which are used both to categorise the link and eventually to retrieve it. So using these keywords, a person can easily seek the link without having to remember the exact name of the site, the title or any of its content. The social aspect of the site implies that bookmarks can be shared amongst different people. This means that anyone can post something interesting to share, however the method of how the sharing occurs is based upon various listings. In fact there are lists which highlight the most recent bookmarks, others which list the most popular, etc. The sharing also means that people tend to share different annotations for the same link, because for one person, a set of annotations might be relevant for a particular link whereas for someone else, a different set might be relevant. The interesting thing is that this techniques serves as a sort of incidental knowledge elicitation whereby users voluntarily add new annotations to the links. However the reason why they add the new knowledge is not to enhance the links but to create a better retrieving mechanism for their needs. Since the byproduct of this process is the annotation of those links, this will

---

[15] http://www.digg.com
[16] http://www.delicious.com

result in the creation of a better tag cloud to represent the link. The positive thing about it is that the more people annotate the link with keywords, the more they manage to refine the tag cloud. Eventually, these tags can be easily used to create folksonomies which represent the link. By taking a wider viewpoint, rather than examining links, a website can be examined as a cloud of links and we can also use the annotation to extract a folksonomy for the site itself. This proofs that we can easily build a powerful system based upon these simple annotations. This power obviously increases as we have more complex tags.

5.3.3   Facebook

The most complex and popular social networking site is probably Facebook[17]. In 2010, the site had more than 500 million active users according to the Facebook statistics[18]. These users would spend more than 700 billion minutes per month on the site. This is not surprising when one considers that the site allows users to:

- Create a personal profile

- Add friends

- Exchange messages and add notifications

- Join groups

- Organise workplace, educational or other information

Apart from being one of the most complex social networking site around, it also has a lot of powerful features related to annotations. These features get their power from the underlying Open Graph protocol[19], which enables web pages representing real world objects to form part of a social graph. These pages can represent a myriad of things from movies, restaurants, personalities, etc. The system allows anyone to add open graph annotations to a web page together with the "Like" button (which is one of the tagging mechanism in Facebook similar to the Digg tagging system described earlier). If a user presses the button, a connection is automatically formed between that page and the user. Subsequently the Facebook programs will gather the information about that page and add the link to the "Likes and Interests" section of the user's profile. So essentially, by adding these features to any website, the site becomes an extension of Facebook. By doing so, the page also appears in other sections of Facebook such as in the search or in the wall thus driving further traffic to that site. In so doing, the owner of the site can get a financial return through adverts placed on the site.

Another important annotation feature on Facebook is photo tagging. Since the site allows users to share photos, it is a common practise for users to annotate the pictures by marking people they know. This ensures that a link is created between

---

[17] http://www.facebook.com

[18] http://www.facebook.com/press/info.php?statistics

[19] http://ogp.me/

the photo and the tagged friend which eventually causes the photo to be displayed on their profile. The tagging process is rather easy, essentially all the users have to do is to click on the face of the person being tagged and a box appears around that face. Even though this process might sound trivial, in essence it is a very powerful approach since it stores:

- The name of the media file (which most of the time is significant)
- The caption underneath the media object
- The exact location of where the photo or video was taken (if it was geo-tagged)
- The people in the media object
- The relationship between the people obtained thanks to the social graph
- The X and Y coordinates of the face pertaining to each and every person in the file
- The site where the document was published

All of these are obtained by simply annotating a document. The interesting thing is that people add the annotations for free simply because of social reasons (i.e. to share the document with other friends). However this social annotation also causes some privacy issues. People can tag anyone in their photos, ever people who prefer not to be on Facebook. So technically, even if a person chooses not to take part in these social networks, there's nothing really stopping his friends from posting his personal details online. Obviously, this can happen with all media however it is much more easier with Facebook since users do not need to learn any particular web language to post online. Apart from these, Facebook allows users to add blogs or micro-blogs on each and every element in the site.

The strength of this system is evident however it still needs to be exploited. Just by considering the photo tagging, it is immediately evident that an intelligent bot can be easily trained to recognise faces. The dataset collected inside the Facebook databases is unprecedented and its potential still needs to be explored.

5.3.4 Flickr

The online photo sharing site Flickr[20] allows users to upload pictures from the desktop, through email or even directly from the camera phone. These pictures are then organised by the user into collections ready for sharing either with anyone around the world or just with a selected few. Flickr also allows users to tag and annotate images.

Tags are essentially labels used to describe the content of a photo. Their primary role is to index the photo thus making it easier for the user to retrieve it. Flickr allows up to 75 tags per photo and different users can only tag a specific photo if they have

---

[20] http://www.flickr.com/

the right to do so. It is interesting to note that this social site is also generating a vocabulary of tags which people can use. This vocabulary includes tags such as:

- photo: images taken by a photographic camera.

- landscape: outdoor images.

- animal: an image of an animal.

- me: a self portrait.

This list is obviously non exhaustive and it is definitely not mandatory however these conventions are helping to bring some order to the Flickr databases. Another important tag is the machine tag which is a normal tag understandable by machines. In fact, what really changes is the syntax used in the tag. These tags are based upon the idea of triples whereby a tag is made up of a namespace, a predicate and a value. These tags are very similar to the conventions mentioned earlier, what really distinguishes them is the namespace. If we take a GeoTag (which binds a picture to a physical location) as an example, this would be written as follows *geo:locality="Rome"*. geo is the namespace, locality is the predicate and Rome is the value. Since the namespace and the locality are fixed (considering they follow the machine tag syntax), programs can be written to parse that tag and understand it. So in this case, the system can easily understand that the picture was taken in Rome. This is very similar to what Twitter is trying to achieve. Eventually, we might see a convergence between these different vocabularies and different programs might be written to understand tags irrespective of whether they originate from Flickr, Twitter or any other system which abides to this structure.

Flickr annotations allow the users to add information to parts of the picture or photo. This is done by selecting an area with the mouse and writing the text associated with the annotation. Since the text entered is essentially HTML, it can also accept hyperlinks. Although this system is very similar to other sites such as Facebook, the fact that it gives the users the liberty to tag photos with any annotations they like (rather than just person annotations) creates new possibilities. As an example, Flickr has been widely used by history of arts lecturers to help students discover new elements in a picture. A typical example can be found in the picture of the Merode Altarpiece[21]. The picture contain about 22 annotations highlighting different aspects such as the symbolism used, the perspective of the picture, the people in the picture, the architecture, the colours utilised, the hidden details and the geometric proportions. The degree of information which annotations can add to a picture is something incredible and technically they are only limited to the annotator's imagination.

The hyperlinks in the annotations provide for further interactivity. A user can easily zoom in a portion of the photo by clicking on the annotations and in so doing, discovering a whole new world of detail. The fact that a picture on Flickr can allow other users to insert their own notes in an image means that a sort of dialogue is created between the owner of the photo and the person viewing it.

---

[21] http://www.flickr.com/photos/ha112/901660/

5.3.5 Diigo

The Diigo[22] application claims to be a personal information management system. In itself it cannot be considered as being a social networking site, however, it allows users to share their annotations thus providing some social features.

Diigo provide users with a browser add-on which allows users to tag any piece of information they find online. What's interesting is the coverage of the application. Since it is an extension rather than a web application, it has to be installed on the different devices. In fact Diigo is available for most of the top browsers including Internet Explorer, Chrome, Firefox, Safari and Opera. Furthermore, it can be installed on Android phones, iPhone and iPad. Data can be imported from sites such as Delicious, Twitter, etc and it can be posted on various sites such as Facebook, Google, Yahoo, etc. Its is this interoperability amongst different services that makes Diigo an invaluable tool. In effect, the system acts as a middle man which provides users with their annotations irrespective of where they are located and independent of the device they are using. This is achieved by making use of the cloud, a remote location where all the annotation is stored.

The level of annotation is very complex offering a myriad of different options including:

- Bookmarks, which allow users to bookmark a page thus allowing them to organise a set of pages in a logical group which makes them much more easier to retrieve at a later stage.

- Digital highlights, capable of selecting pieces of texts from any site. This text can also be colour coded in order to assign specific categories to the text.

- Interactive stickynotes provide the possibility of adding whole notes to a particular area in a website. Essentially, this feature is very similar to what is normally found in modern word processors whereby a whole block of text in the form of a note can be attached to a specific slot in the document.

- Archiving allows for whole pages or snippets of the page (stored as images) to be recorded for an indefinite amount of time. These pages can also be annotated using markers (since the object being manipulated is an image). Apart from this, keywords can also be assigned to these pages in order to make them searchable.

- Tagging provides users with the facility to add keywords to a specific page or snippet. This makes it easier to locate and retrieve.

- Lists are logical collections such as bookmarks which can also be ordered. In fact, apart from providing membership, a list can also allow the users to organise its elements and eventually even present them in a slideshow.

The system also supports sharing. A number of different privacy options are available to the users whereby an annotation can be public or private. These annotations

---

[22] http://www.diigo.com/

can also be curated by a group of users thus changing the static pages into live views which evolve with time.

5.3.6 MyExperiment

The social collaborative environment called MyExperiment[23] defined in [102] [101] [73] [188] [171] and [3] allows scientists to share and publish their experiments. In essence this forms part of a different breed of social networking sites which are specialised in a particular domain. These sites can be considered as a formalisation of Communities of Practise (COP). Whereas before, the collaboration between different people sharing a common profession was haphazard, sites such as MyExperiment managed to consolidate everything inside a social networking site.

The aim of the site is multifaceted. First and foremost, it aims to create a pool of scientific knowledge which is both accessible and shared between the major scientific minds. Through this sharing, it promotes the building of new scientific communities and relationships between individual researchers. It also promotes the reuse of scientific workflows thus helping scientists reduce time when designing experiments (since they would be using tried and tested methods which avoid reinventing the wheel).

In the case of this web application, rather than having images or documents as in most other websites, the elements annotated are actually workflows. A workflow is essentially a protocol used to specify a scientific process in this case. The application which is based on the Taverna Workflow Management System[24] ensures that workflows are well defined and provides features to share the workflows thus making them easier to reuse. By doing so, if a scientist needs to create a similar process, it is simply a matter of finding the workflow, modifying it to suite its needs and applying the process. By reusing these processes, scientist would be avoiding errors thus making it quicker for them to test their ideas. Without such a system, the reuse of workflows would be incredibly cumbersome. Individuals or small groups working independently of each other or in distant geographic locations would find it problematic to interact together. There might be processes that go beyond the expertise of the person or the group thus the social element comes into play. In some cases, the process even crosses amongst different disciplines thus new blood would have to enticed in order to enhance the working group.

Apart from the normal tagging and micro-blogging associated with social networking sites, the system also allows users to manage versions and licencing, add reviews, credits, citations and ratings. Versioning and licencing are extremely important when dealing with high reusable components. The fact that metadata is added to the workflow in order to store this additional information enhances its use. Reviews are rather different than the micro-blogs. In essence, they have a similar format however semantically, the scope of a review is to evaluate the whole process. On the other hand, micro-blogs can focus on a particular part of the workflow and

---

[23] http://wiki.myexperiment.org/
[24] http://www.taverna.org.uk/

not consider it in its entirety. Credits are used to associate people to a workflow, they identify who created or contributed to the creation of the process and to what degree. Once again, myExperiment is focusing on the social element behind these processes. Citations, are inbound links from the publications to the workflow. These links are not only used to annotate a workflow with metadata but they also ground the process to sound scientific experiments that were published in various domains. Finally, a rating allows users to vote for their favourite process thus serving as a recomendation to others intending to use the workflow.

5.3.7 Twitter

The social networking site Twitter[25] offers users the facility to post micro-blogs called tweets. Users can follow other people and subscribe to a feed which is updated every time a tweet is posted. In recent months, Twitter also added the possibility of having annotations.

The system allows users to add various annotations to a single tweet using structured metadata. For Twitter, annotations are nothing more than triples made up of a namespace, a key and a value. These annotations are specified when the tweet is created and an overall limit on the size of the annotation is imposed by the company. The system is quite flexible and users can add as much annotations as they like. The type of data which can be added to the annotations is restricted by XML since it is used as the underlying format. However one can easily surpass this restriction since rather than attaching a binary file, a user can always place the file somewhere online and attach the URL to that file. Another property of these annotations is immutability. This means that if a tweet has been published with annotations, the user or the author cannot change them. Notwithstanding this, one can always retweet posting and in that case, new annotations can be added.

The uses of such a system are various and they are only restricted by the user's needs or imaginations. The following is a non-exhaustive list of some of these uses:

- Rich media ranging from presentations to movies can be included as links in the annotations.
- Tweets could have a geo-location associated to them thus giving them a third dimension. In this case, a tweet could simply be a comment attached to a physical building.
- Advertisers can use this technology to add related stuff to a tweet.
- Sorting and searching might be enhanced using keywords in the annotations.
- Feedback from users obtained through blogs or surveys might be associated to a tweet.
- Social gaming.

---

[25] http://twitter.com/

- Used to connect a tweet to other chat clients.

- Posting a tweet in multiple languages.

- Share snippets of codes or bookmarklets.

The only problem so far is that there isn't really any standard for annotations in Twitter. This might lead to compatibility issues related to metadata.

5.3.8 *YouTube*

The video sharing site YouTube[26] allows people to post video clips online and share them. The site provides similar social features as other sites such as the "like" button, micro-bloging, the possibility to share movies and also to subscribe to particular channels. Two notably differences in youTube are the "don't like" button and the advanced video annotations.

The "don't like" button is similar to the "like" button but rather than posting a positive vote, it posts a negative one. Such posts having negative connotations are not widely spread in social sites. In fact, such a button is absent from the major social sites. The idea of having an annotation with positive and negative connotations is a reflection of the democratic nature of the system. Such a system implies that a media file (in this case a video) posted by someone does not gain popularity simply because a lot of people like it, but the person posting it needs to be careful that a lot of people do not dislike it. Thus, it offers a fair perspective of the video's value. However it is obvious that this notion does not apply to anything which can be annotated. If the user is annotating a personal photo, it doesn't really matter if someone else dislikes it because the user is sharing it for social reasons and not to take part in a contest. Also, when it comes to artistic media, the liking of an artefact is subjective to the person viewing it and there is no rule cast in stone which defines what is aesthetically pleasing or not.

The other annotational features of YouTube are rather advanced. Given a video, the system allows the users to add five different kind of annotations:

Speech Bubble can be added to any part of the video. They will pop-up in the specified location and remain visible for a predefined period of time. These bubbles contain text inside them and are normally used in conjunction with movies of people, animals or even objects expressing their opinion through the speech bubbles.

Spotlight allows users to select a portion of screen which needs to be highlighted during the viewing of the video. This is achieved by showing a box with a thin boarder around the area. Users can also add some text around the box.

Notes are similar to Speech Bubble but they have a different shape (just a square box) and they do not have a pointer. However, the functionality is exactly the same.

---

[26] http://www.youtube.com/

Pause allows the user to freeze the video for a specified period of time.

Title creates a piece of text which can be used to add a title to the video.

These annotations essentially server to provide additional information to the person watching the video and they allow users to link to other parts of the web. The latter can be added to most of the annotations mentioned above. This linking also provides for some degree of interactivity since users can be presented with a choice and they can make their choice by simply selecting a link out of a group of possible options.

However, this system still has some open issues. The editing options provided by YouTube are very limited in fact users can't copy or paste annotations. They are not stored as indexable metadata thus they are not indexed by the major search engines. Even though users can change the colours associated with an annotation, since annotations have a life span, the annotation might be hard to spot when the background image changes. Notwithstanding these issues, YouTube still offers a powerful annotation tool which provides users with a new experience when sharing videos.

5.4   Conclusion

In this chapter, manual annotation was shown under a new light, one which uses the power of Web 2.0 technologies such as Social Networking and internet applications which are both useful as in the case of the reCAPTCHA and entertaining as in the other examples. The notion of having several humans working manually on such complex tasks was unthinkable until a few years ago, however, today it seems that these approaches are making human collaboration possible. In the coming chapters, Artificial Intelligence will further help in the annotation process thus reducing the dependency on humans.

Chapter 6
Semi-automated Annotation

The various approaches described so far are effective for controlled tasks such as annotating a collection of patient records. In reality, very few of these techniques really scale effectively to produce an ongoing stream of annotations. However, every controlled task is problematic. We live in a dynamic world where things constantly change and probably those annotations would have to change with time. The patient record would have to be updated, patients die and new ones are recorded. So rather than inserting the annotations in the records, the best approach would be to create an ontology and insert in the ontologies a reference to the instances rather than the actual instances. In this way, if the instance changes slightly, there is no need of modifying all the ontologies where this instance appears since the link would still be valid. This also makes sense because in our world and even on the Internet, there exists no Oracle of Delphi [74] that has the answers to all the possible questions thus we can never be sure of the validity of our data. Knowledge is by nature distributed and dynamic, and the most plausible scenario in the future [108] seems to be made up of several distributed ontologies which share concepts between them. This document already delved into the issues why the annotation task might be difficult when performed by humans. If we think about the current size and growth of the web [62], it is already an unmanageable process to manually annotate all of those pages. If we re-dimension our expectations and try to annotate just the newly created documents, it is still a slow time-consuming process that involves high costs. Due to these problems, it is vital to create methodologies that help users during the annotation of these documents in a semiautomatic way.

6.1 Information Extraction to the Rescue

One of the most promising technologies in the **HLT!** (**HLT!**) field, is without doubt Information Extraction (IE). IE is a technology used to automatically identify important facts in a document. The extracted facts can then be used to insert annotations in the document or to populate a knowledge base. IE can be used to support in a semi/automatic way knowledge identification and extraction from web documents (E.g. by highlighting the information in the documents). Also, when IE is

A. Dingli: Knowledge Annotation: Making Implicit Knowledge Explicit, ISRL 16, pp. 59–69.
springerlink.com   © Springer-Verlag Berlin Heidelberg 2011

combined with other techniques such as Machine Learning (ML), it can be used to port systems to new applications/domains without requiring any complex settings. This combination of IE and ML is normally called Adaptive IE [152][2][1]. It has been proven in [216] that in some cases, these approaches can reduce the burden of manual annotation up to 80%. The following section, will have a look at the various semi-automatic annotation tools.

6.1.1   The Alembic Workbench

The Alembic Workbench [70] is one of the first systems created out of a set of integrated tools that make use of several strategies in order to bootstrap the annotation process. The idea behind Alembic is that when a user starts annotating, the inserted annotations are normally not only bound to that specific document but they can also apply to other similar documents. This interesting observation can be used to reduce the annotation burden by reusing these annotations in other documents. Therefore in Alembic, every piece of information which can be used to help the user is utilised. Eventually, when the user is confident with the accuracy of the system, the task of the user changes from one of manual annotator to one of manual reviewer. The annotations are inserted into Alembic by marking elements using a mouse. Together with this method of manual annotation, some other strategies are used in order to facilitate annotation such as;

- String matching algorithms ensure that additional instances of marked entities are found throughout the document.

- Built-in rule languages are used to specify domain specific rules which are then used for tagging.

- A pattern system is used to mine for potential phrases and suggest possible patterns to the user.

- Statistical information which identifies important phrases, frequency counts, etc provide users with important information.

The most innovative feature of this application is the use of pre-tagging. The main idea is that information which can be identified before the user starts tagging should be tagged in order to support the user by preventing him from wasting time on trivial elements which can be tagged automatically by the system.

Another innovative feature of Alembic is the implementation of a bootstrapping strategy. In this approach, a user is asked to mark some initial examples as a seed for the whole process. These examples are sent to a learning algorithm that generates new examples and the cycle repeats. Eventually a number of markings are obtained and are presented to the user for review. If the user notices that some of the rules are generating incorrect results, it is possible for the user to manually change the rules or their order so that the precision of the algorithm is increased. Although the machine learning rules generate quite good results, they lack two important factors which humans have i.e. linguistic intuition and world knowledge. The Alembic methodology does not cater for redundant information, therefore allowing documents which are already covered by the IE system to be presented to the user for annotation. This makes the annotation process more tedious and time consuming for the user. Experiments performed using the Alembic workbench has showed significant improvements in the annotation of documents. In several tests, it was shown that users double their productivity rate. Also, with the data provided both by the users and automatically from the system, it was possible to train quite complex IE tools.

6.1.2   The Gate Annotation Tool

The General Architecture for Text Engineering (GATE) [66][33] is an infrastructure which facilitates the development and deploying of software components used mainly for natural language processing. The packages come complete with a number of software components such as IE engines, Part of Speech taggers, etc and new components can be added quite easily. One of the main features of the Graphical User Interface (GUI) provided with GATE is the annotation tool. The annotation tool is first of all an advanced text viewer compliant with many standard formats. A document in GATE is made up of content, annotations and features (attributes related to the document). The annotations in GATE (as any other piece of information) is described in terms of an attribute-value pair. The attribute is a textual description of the object while the value can represent any java object (ranging from a simple annotation to a whole java object). These annotations are typed and are considered by the system as directed acyclic graphs having a start and end position. The type depends on the application, they can be atomic such as numbers, words, etc but they can also be semantically typed referring to a person, an institution, a country, etc. This is possible thanks to another IE engine found in gate called ANNIE [155].

The annotation interface works like similar tools whereby a user selects a concept from an ontology and highlights the instances of the concept in the document. The system also supplies some generic tools which are capable of extracting generic concepts from documents. These tools can also be extended by using a simple grammar to cover more domain specific concepts. Being an architecture, GATE allows other external components to be loaded which can aid to locate concepts. The results of these tools are then presented in the annotation interface in the form of a tree of concepts. The user simply needs to select a concept or a group of them and the annotations are immediately displayed in the document viewer as colored highlights. The GATE annotation tool is a powerful tool since it allows several independent different components to work together seamlessly. It also presents the user with a unified view of the results which were obtained from the different components.

6.1.3   MnM

[81] describes an annotation tool called MnM that aids the user in the annotation process by providing semi-automatic support for annotation. The tool has integrated

in it both an ontology editor and a web browser. MnM support five main activities browse, markup, learn, test and extract.

- Browsing is the activity of presenting to the user ontologies stored in a different location through a unified front-end. The purpose of this activity is to allow the user to select concepts from the ontologies which are then used to annotate the documents in future stages. To do so, the application provides various previews of the ontologies and their data. This part is also referred to as ontology browsing.

- Markup or Annotation is done in the traditional way, i.e. by selecting concepts from the chosen ontology and marking the related text in the current document. This has the effect of inserting XML tags in the body of the document in order to semantically mark specific sections of the document.

- For the learning phase, MnM has a simple interface through which several learning algorithms can be used. The IE engines tested were various ranging from BADGER [94] to Amilcare [59][58]. The IE engine is used to learn mappings between annotations in the documents and concepts in the various ontologies.

- With regards to the testing, there are basically two ways, explicit or implicit. In the explicit approach, the user is asked to select a test corpus which is either stored locally or somewhere online and the system performs tests on that document. In the implicit approach, the user is still asked to select a corpus like the implicit approach but the strategy for testing is handled by MnM and not all the documents are necessary used for testing.

- The final phase is the extraction phase. After the IE algorithm is trained, it is used on a set of untagged documents in order to extract new information. The information extracted is first verified by the user and then sent to the ontology server to populate the different ontologies.

MnM is one of the first tools integrating ontology editors with an annotation interface. Together with the support of IE engines, these approaches facilitate the annotation task thus relieving most of the load from the users.

6.1.4   S-CREAM

Another annotation framework which can be trained on specific domains is S-CREAM [116][117] (Semi-automatic CREAtion of Metadata). On top of this framework, there is Ont-O-Mat[114], an annotation tool. This tool makes use of the Adaptive IE engine AMILCARE. AMILCARE is trained on test documents in order to learn information extraction rules. The IE engine is then used to support the users of Ont-O-Mat, therefore making the annotation process semi-automatic.

The system once again makes use of an Ontology together with annotations. In this application, annotations are elements inserted in a document which can be of three types; tags part of the DAML+OIL domain, attribute tags that specify the type of a particular element in a document or a relationship tag. A user can interact with the system in three ways;

- by changing the ontology and the templates describing facts manually,

- by annotating the document and associating those annotations with concepts in the ontology,

- or by selecting concepts from the ontology and marking them in the document.

After the initial annotations provided by the user, S-CREAM exploits the power of adaptive IE to learn how to automatically annotate the document. Obviously, this can only occur after the IE engine is trained on a substantial number of examples provided by the user. The last kind of process uses a discourse representation to map from the tagged document to the ontology. This discourse representation is a very light implementation of the original theory. The reason being that discourse representation was never intended for semi-structured text but rather for free text. Therefore to overcome this limitation, the one used in S-CREAM is a light version made up of manually written logical rules in order to map the concepts from the document to the ontology.

S-CREAM is a comprehensive framework for creating metadata together with relations in order to semantically markup documents. The addition of an IE engine makes this process even easier and helps pave the way forward towards building automated annotation systems.

6.1.5 Melita

Melita [56][55] is an ontology-based text annotator similar to MnM and S-Cream. However, the major difference is that at the basis of the system, there are two user-centred criteria: timeliness and intrusiveness of the IE process. The first refers to the time lag between the moment in which annotations are inserted by the user and the moment in which they are learnt by the IE system. In systems like MnM and Ont-o-mat this happens sequentially in a batch. The Melita system implements an intelligent scheduling in order to keep timeliness to the minimum without increasing intrusiveness. Thus, the system does not take away the processing power which might be required by the user and in fact, the user is unaware that Melita is learning in the background whilst he's continuing with his manual annotations. The intrusiveness aspect refers to the several ways in which the IE system gives suggestions to the user without imposing anything on the user.

In Melita, the annotation process is split into two main phases; training and active annotation with revision. In user terms, the first corresponds to unassisted annotation, while the latter mainly requires correction of annotations proposed by the IE engine.

While the system is in training mode, the system behaves in a similar way to other annotation tools. In fact, at this stage, the IE system is not contributing in any way to the annotation process. However, the devil is in the details and even though the user is not noticing anything, if we take a closer look to what is actually happening in the background, we find that the system is not dormant. The IE uses the examples supplied by the user to silently learn and induce new rules. This phase can be referred to as the bootstrapping phase whereby the user supplies some seed examples for an arbitrary document. The system then learns new rules that cover those examples. As soon as the user annotates a new document, the system also annotates the document using the rules it learnt previously, and compares its results with those of the user. In this way, the system is capable of evaluating itself (when compared with the user). Missing annotations or mistakes are used by the learning algorithm to learn new rules and adjust existing ones. The cycle continues like that until the system reaches a sufficient level of accuracy predefined by the user (Different levels of accuracy might be required for different tasks). Once this level is reached, the system moves over to the phase of active annotation with revision.

In this phase, Melita presents to the user a previously unseen document with annotations suggested by the system itself. At this stage, the user's task shifts from one of annotator to one of supervisor. In fact, the user is only expected to correct and integrate the suggested annotations (i.e. removing wrong annotations and adding missing ones). When the document is corrected, these are sent back to the IE system for retraining. By applying corrections, the user is implicitly giving back to the system important feedback regarding its annotation capabilities. These are then used by the system to learn new accurate rules and therefore improve its performance.

The task of the user is also much lighter than before. Supervising and correcting the system is much easier and less error prone than looking for instances of a concept in a document. It is also less time consuming since the attention of the user is mainly focused towards the suggestions given by the system and the need of new manual annotations decreases when the accuracy of the IE system increases.

6.1.6   LabelMe

[191][192] created an annotation tool called LabelMe which specialises on image annotation. To do so, they make use of similar techniques mentioned in Section 5.2.1 whereby various users collaborate online to annotate a database of images. The annotation is quite powerful and allows users to not only assign keywords to an image but also to annotate specific objects in the image by drawing a border around those objects and associating annotations to it. However, the distinguishing factor that sets it apart from the applications mentioned in Chapter 5 is that it can annotate the images semi-automatically.

The process adopted by LabelMe is similar to what we have seen already. Essentially, a set of images is manually annotated, a classifier is then used to learn the boundaries of the annotations associated to a particular image and the trained classifier is then used to identify objects in previously unseen images. To further support the annotation process, WordNet[1] is used. Essentially WordNet is a large dictionary of English words containing meanings and relationships between the words. By using these relationships, the system can suggest sub components of objects found in the image thus facilitating the annotation task.

---

[1] http://wordnet.princeton.edu

6.2 Annotation Complexity

Most of the algorithms mentioned so far provide quite a substantial improvement in some cases even saving the user up to 80% of the annotation process. However this is not enough since different concepts differ, some might be easier to spot whilst others might be extremely complex. To explain this annotational complexity, we can have a look at various data sets and examine why some annotations are more difficult than others.

Of particular interest is the Carnegie Mellon University (CMU) seminar announcements corpus. Essentially, this is a corpus widely used in IE ([96], [44], [52]) and considered by many as being one of the gold standards in the field. The CMU seminar announcements corpus, consists of 485 documents which were posted to an electronic bulletin board at CMU. Each document announces an upcoming seminar organised in the Department of Computer Science. The documents contain semi-structured texts consisting of meta information like the sender of the message, the time of the seminar, etc together with free text specifying the nature of the event. The idea behind this domain is to train an intelligent agent capable of reading the announcements, extract the information from them and if the agent considers the seminars to be relevant (based upon some predefined criteria) they are inserted directly in the user's electronic diary.

The fields extracted for the task include:

Speaker the full name including the title of the person giving the seminar.

Location the name of the room or building where the seminar is going to be held.

Start Time the starting time of the seminar.

End Time the finishing time of the seminar.

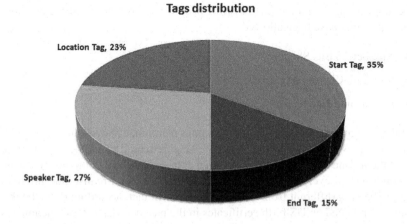

Fig. 6.1 Distribution of Tags in the CMU seminar announcement corpus

The tags in the corpus are not distributed equally as can be seen in Figure 6.1. It seems that the corpus is very rich in Start Time tags and less rich in End Time tags. This is not surprising since in general, people are more interested in the start of a seminar than in its end. Also, the end of a seminar may be fuzzy since events normally take longer than expected especially when there's a question and answering session towards the end of it. The location and speaker tags can be found in almost similar amounts since each seminar would have at least one speaker and one location. Just by examining this information, we might probably deduce that Start time will be easier to extract because there are a lot of instances whereas End time will be difficult to extract because there are fewer instances. However this assumption does not hold. Apart from the distribution of tags, one has to consider the nature of the tags and also its representation within the corpus.

An examination of the four tags mentioned earlier will help us understand this issue:

Speaker will be quite difficult to learn. Intuitively, we know that the name of a Speaker can be any sequence of letters. A named entity recogniser (such as [155]) can help spot names using linguistic cues (such as the title Mr, Mrs, Dr, etc before a word). In fact, these systems have powerful grammars which can be customised to spot these cues, however, these cues are not always found within the text. Apart from these rules, such a system would make extensive use of gazetteers to spot named entities belonging to specific semantic groups. These work by using huge lists of names harvested from the web. Sometimes, these lists might include thousands of names, however this approach has its problems as well. Really and truly, what constitutes a name is up to the people who gave that name to that person. Normally, people choose common well known names, however this is not always the case. On the 24th July 2008, the BBC reported that Judge Rob Murfitt from New Zealand allowed a nine-year old girl to change her name because it could expose her to teasing. However he also commented that the public registry should be more strict when it comes to naming people and gave the following examples of names that have been allowed:

- Number 16 Bus Shelter

- Midnight Chardonnay

- Benson and Hedges

If the recogniser tries to figure out the names mentioned above without any additional linguistic cues and just relying on the gazetteer, it would be impossible to find them. One might argue that this is an extreme case and definitely not within the norm. However, if we have a look at [150] we soon realise that the norm might be very different from what one expects. In fact, according to the book, a study conducted on US birth certificates in the past decades shows that amongst the most popular names, one can find:

- Holly

- Asia

- Diamond

The name of a shrub, of a continent and of a precious stone were gradually adopted by people to name their children. Since these words are used as labels to refer to both objects and persons, it becomes increasingly hard for an automated system to determine in which context they are being used. They cannot be added to the gazetteer because they would automatically introduce noise and the only way to reliably detect them is through the fabrication of complex detection rules. There are also other cultural implications one should consider. Western and eastern names are very different from each other. Thus different gazetteers need to be used to extract people's names. However in a multicultural society this solution is not feasible and a gazetteer of some 40,000 names would easily not suffice to cover the various names encountered.

The task at hand in the seminar announcements is even more complex because it is not simply a matter of spotting names but the system must identify the person who is going to give the seminar. So if we find two names in a document, one identifying the host person and the other identifying the speaker, the system must only return the name of the speaker and discard that of the host (Even though it is correctly recognised as a person's name). From the data in Figure 6.2 we can see that there are around 491 distinct phrases containing names meaning that there is more than 1 new phrase (containing a name) per document. Even though there are 757 examples (around 27% of all tags) in the documents containing the Speaker tag, the fact that many of these examples are new and not repeated elsewhere, makes the whole task much harder.

Location is yet another difficult category. The name of a geographical location can be anything, just as a person's name. To complicate things further, it can also be used to name a person, thus adding further confusion. Another problem with locations is that they are not unique and this creates a problem with Semantic Taggers. If the tagger tries to identify the country pertaining to a particular location, say Oxford, it will be hard for it to establish the correct one. In Europe, there is just one place named Oxford which is located in the United Kingdom (UK). However, in the US, there are up to four places named Oxford; in Ohio, Mississippi, Alabama and Michigan. To disambiguate further, one has to gather more contextual information from the document if it is available.

The complexity of identifying such an entry changes primarily based upon the domain. An open world domain where the information can refer to any geographical point around the world or even beyond, is definitely much more complex. Gazetteers do help in this case; they still suffer from some problems mentioned earlier however, if they are verified with online sources, they can produce very good results. Sites such as WikiTravel[2] and VirtualTourist[3] record informaton about 50,000 locations each. If we analyse these locations, we'll immediately realise that they are the most popular locations around the planet. If the document refers to some less known location, then the Gazetteer might face some problems and the only way to spot the location would be via linguistic cues. Closed domains on the other hand normally refer to fewer entries which form part of a specific grouping. This grouping varies and there can be various reasons; it can range from being a set of geographical locations mentioned in a novel (such as The Da Vinci Code Trail[4]) so in this case, the linkage between the locations in the domain is purely fictitious. Or it can simply be a case of geographical proximity such as in the CMU seminar announcements. In either case, a generic gazetteer would be of little use and the approach to identify these locations either requires the handcrafting of specialised gazetteers or the creation of specific rules. The reason for this being that it would be highly unlikely to find a list of these specific locations somewhere online.

If we try to handcraft the rules, an analysis of the corpus reveals that the total number of examples in the corpus pertaining to a Location amounts to 643 or 23% of all the tags. Out of these, the total number of distinct phrases is 243 which means that slightly more than $\frac{1}{3}$rd of the tags are new. This proportion is quite significant when considering that the corpus is based on a closed domain. In this case, the learner needs to single out these unique locations and learn patterns in order to identify them.

Start Time on the other hand is a completely different story. A temporal element can have various forms, it can be expressed both in numeric (Eg. 13:00) or textual form (Eg. One O'clock). However, even though there are different representations for the same time, the different ways of expressing it is quite limited. Further still, a semantic tagger capable of identifying time can be easily used across different domains. In fact, we can see that the rules for Start Time can easily apply for End time as well. However there are still a few challenges which need to be overcome. When we have two or more temporal elements in the same documents such as in this case (start time and end time), we need to take a slightly different approach. First of all, the system needs to identify the temporal elements and this is done by using a semantic tagger. The next step is to disambiguate between the different types. To achieve this, the context of the elements is used. The documents being processed contain a total of 982 tags (or 35% of the total number of tags) and there are only around 151 distinct phrases. This means that training on 15% of all the documents (almost 75 documents) is enough to learn this concept. End Time is slightly more complex. The number of distinct phrases are very little, around 93 instances, but so is the representation of this concept in the corpus. In fact, this concept appears only around 433 times (i.e. 15% of all the tags). This means that there is a substantial number of documents where

---

[2] http://http://wikitravel.org

[3] http://http://www.virtualtourist.com/

[4] http://www.parismuse.com/about/news/press-release-trail.shtml

the End Time concept is not represented. So start time is relatively easier to learn since the documents have various examples most of which are repeated and as a consequence, easier to learn. End time is somewhat more complex however, since we are aware that the corpus contains only two time related tags, we can use logical exclusion to identify the End time. So if the element has been tagged as being a temporal element but the system is not sure if it is an End time or a Start time, the fact that the Start time classifier does not manages to identify it automatically makes it an End time.

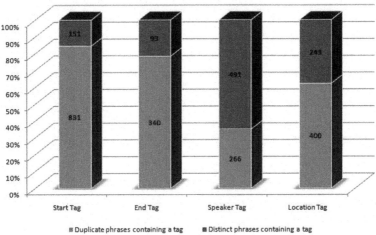

Fig. 6.2 Different phrases containing a tag which were found in the document

6.3 Conclusion

In this chapter, various semi-automated annotation approaches were analysed. In all of them, the improvement provided is quite substantial, in some cases taking over the bulk of the annotation process. However this is not enough since different concepts differ, some might be easier to spot whilst other such as images might be extremely complex. We have seen how both the corpus and the concepts can be examined and their difficulty examined. In the coming chapter, will have a look at how the annotation process can be fully automated thus removing the human element from the loop.

Chapter 7
Fully-automated Annotation

Even though semi-automatic annotation is a huge step forward from manual annotation, it is still based on human centred annotation. Although the methodology relieves some of the annotation burdens from the user, the process is still difficult, time consuming and expensive. Apart from this, considering that the web is such a huge domain, convincing millions of users to annotate documents is almost impossible since it would require an ongoing world-wide effort of gigantic proportions. If for a second we assume that this task can be achieved, we are still faced with a number of open issues.

In the methodologies we've seen in Chapter 6, annotation is meant mainly to be statically associated to (and in some cases saved within) the documents. Static annotation associated to a document can:

1. be incomplete;

2. be incorrect (when the annotator is not skilled enough);

3. become obsolete (not aligned to page updates);

4. be irrelevant for some users since a different ontology can be applied to the document (Eg a page about flowers might have annotations related to botany, caring for plants, medical properties of the flower, etc).

Web 2.0 applications are already hinting to the fact that in the near future, most of the annotations would be inserted by Web actors other than the page's owner, exactly like nowadays, search engines produce indexes without modifying the code of the page. Producing methodologies for automatic annotation of pages with minimal or no user intervention becomes therefore important. Once this is achieved, the task of inserting annotations loses its importance since at any time, it would be possible to automatically (re)annotate the document and to store the annotation in separate databases or ontologies.

Because of these needs, in the coming sections, we'll have a look at various methodologies which learn how to annotate semantically consistent portions of the

A. Dingli: Knowledge Annotation: Making Implicit Knowledge Explicit, ISRL 16, pp. 71–77.
springerlink.com   © Springer-Verlag Berlin Heidelberg 2011

web. All of the annotations are produced automatically with almost no user intervention, apart some corrections which the users might want to perform.

7.1 DIPRE

Dual Iterative Pattern Relation Expansion (DIPRE) is the system defined in [41] whose job is to extract information from the web in order to populate a database. By doing so, annotations can be created as a byproduct of the extraction process since the data extracted from the documents can be easily marked in the document. The interesting aspect of this process is that the system works almost automatically. The user initiates the process by simply defining a few seed examples which are used to specify what is required by the algorithm. In the experiments provided, five examples were enough to bootstrap the process of finding authors and their books. The system makes use of the initial examples to generate patterns. These patterns are made up of essentially five elements; the prefix, the author, the middle section, the book title and the suffix. Essentially the prefix, middle section and suffix are regular expressions generated automatically.

The process is as follows; occurrences of the seed examples are sought by using a search engine. The result is a collection of different documents containing instances of those examples. A learning algorithm is then used to generate patterns by using the collection of documents harvested from the web. The patterns generated are then applied to new web pages in order to discover more instances. The process continues until the system stops generating new instances. The job of the user is simply to monitor the system and correct any erroneous patterns.

The documented results are very interesting, first of all, the 5 initial seed examples generated a total of 15,000 new entries. This is quite impressive when considering the amount of work required by the user when compared to all those instances generated. Secondly, the error rate was about 3% of the entries which is quite good especially with respect to the large amount of correct entries. Finally, even though at the time, Amazon claimed to catalogue around 2.5 million books the algorithm still managed to find books which were not represented in the collection. This reflects a lot on the nature of the web, its redundancy and the nature of distributed information. The downside of this approach was two fold, first of all the system was rather slow since it had to go through millions of pages. Secondly, the instances identified in 5 million web pages amounted to less than 4000 instances which is quite low and the cause of this merits further studies.

7.2 Extracting Using ML

Tom Mitchell (the author of various ML books including [165]) describes his work at WhizBang Labs in [166] where he applied ML to IE from the web. The idea is to automatically annotate entities found online which include dates, cities, countries, persons, etc. If these elements are semantically annotated, the process of identifying information in web pages via search engines would be much more accurate. To extract this semantic information, he uses three types of algorithms;

- The Naive Bayes model[1] is used to automatically classify documents based upon particular topics (which are identified automatically through keyword analysis).

- An improvement on the previous model is the "maximum entropy" algorithm, which go beyond independent words and examine the frequency of small phrases or word combinations.

- The last approach is called co-training which examines hyperlinks in the page and associates with them keywords from the document they refer to.

In synthesis, the important thing about these approaches is that they use generic patterns together with ML techniques in order to improve the extraction process.

7.3 Armadillo

The Armadillo methodology [47][53][78][54] exploits the redundancy of the web in order to bootstrap the annotation process. The underlying idea was inspired from [166] and [41] however, the novelty behind Armadillo is that it uses both approaches concurrently whilst exploiting the redundancy of information available online. Since redundant information might be located on different web sites, the system also implements various Information Integration (II) techniques to store the information extracted into one coherent database.

The methodology is made up of three main items which include a set of Strategies, a group of Oracles and a set of Ontologies or Databases where to store the information.

Strategies are modules capable of extracting information from a given document using very simple techniques. Each strategy takes as input a document, performs a simple extraction function over that document and returns an annotated document. The input document is not restricted to any particular text type and can range from free text to structured text. It can also be extended to annotate pictures or other media types quite easily. The extraction functions found in the strategies use rather weak techniques such as simple pattern matching routines. The idea is that whenever weak strategies are combined together, they manage to produce stronger strategies. To better illustrate the role of a strategy, imagine a system whose task is to extract the names of authors. A very simple yet highly effective heuristic would be to extract all the bigrams[2] containing words starting with a capital letter. This works pretty well and manages to return bigrams like "Tom Smith", etc. One can argue that this approach would probably return some garbage as well like the words "The System". This is true, but this problem is solved in two ways; first of all, a

---

[1] A probabilistic classifier which assumes that the presence or absence of a feature is independent from the presence or absence of another feature.

[2] A bigram refers to phrases made out of two words.

postprocessing procedure inside it is used to filter away garbage by using simple approaches (such as removing bigrams containing stop words[3]) and secondly, before annotating elements in texts, Armadillo must verify them by using an Oracle. The most important thing is that these strategies provide some seed examples which are used by the Oracles to discover other instances.

An Oracle is an entity which can be real (such as the user) or artificial (such as a website directory), that possesses some knowledge about the current domain. This adds a certain degree of accountability to the system since an Oracle is responsible for the data it validates. Therefore, if an item of data is wrong and it was validated by a particular Oracle, the system can pinpoint exactly which Oracle validated the data and take appropriate corrective actions. These actions include adjusting the validation mechanism of the Oracle or even excluding it from future validations. The exclusion of an Oracle is normally not a big loss since a system would normally have different Oracles performing the same validations. However these validations would use different methods thus exploiting the redundancy of the web to the full. The combination of these Oracles will produce very reliable data since it is not up to one Oracle to decide if an instance is valid or not but rather to a committee of Oracles (similar to having a panel of experts to evaluate the data). Another task of the Oracle is to augment any information it might posses together with the information being annotated. There can be different types of Oracles such as humans, gazetteers, web resources and learning algorithms.

- Humans are the best kind of Oracles since they posses a huge store of information and they are excellent information processing systems. The problem with humans is that the whole scope of this system is exactly to spare the hassle of inserting annotations. Because of this, this Oracle is mainly used to produce some seed examples and to verify the data.

- Gazetteers are lists of elements which belong to the same class. These lists can contain anything like lists of names, countries, currencies, etc. Each list is associated with a concept found in one or more ontologies. As an example, if the system is processing the words "United Kingdom", a search is performed through the lists to identify whether this word is an instance that occurs in any of the available lists. In this example, the phrase "United Kingdom" was found in the gazetteer called **countries**. This gazetteer is also attached to a concept found in one of the ontologies which is called **country**. Therefore, the system assumes that "United Kingdom" is equivalent to the instance found in the countries gazetteer and thus, it is semantically typed as being a **country**.

- Web resources include any reliable list or database found over the web which can be used to verify whether an instance is part of a particular class or not. The information must be very reliable and up to date since at this stage, a minor error rate is not allowed. The task of querying these web resources is not always straightforward. First of all, if the web resource is accessible through a web service, the system can easily access the web service by making use of standard

---

[3] Stop words are words frequently occurring in texts such as the articles (a, an, the, etc).

techniques. If no web service is available, a strategy must be defined by the user, in order to instruct the system, which information to extract from the web page. Luckily, these pages are normally front ends to databases, therefore, since their content is generated on the fly using some program, the layout of the page would be very regular.

- Learning algorithms include technologies such as ML and IE tools. These algorithms are much more sophisticated than the other approaches. They specialise on annotating information from individual documents which are not regular and therefore, learning a common wrapper as we have seen before is impossible. These algorithms do not even need any training by humans. They typically get a page, partially annotate it with the instances which are available in the database, they learn from those instances and extract information from the same page. The cycle continues like that until no more instances can be learnt from the page.

The Armadillo system proved itself to be very efficient. In fact, a problem of such a system is that it either manages to obtain high precision[4] and low recall[5] or vice-versa. In this system, the information integration algorithms manage to produce few results which are extremely precise. When they are combined with the IE part of the system, they managed to get high precision and high recall which is quite rare for such systems thus emphasising the success of this methodology. These results were repeated on various different domains.

7.4 PANKOW

[51] describes an annotation component called PANKOW (Pattern-based Annotation through Knowledge On the Web) which eventually replaced Ont-O-Mat (described in Section 6.1.4). In spirit, it is very similar to Armadillo, however the interesting aspect of this system is that it generates seed elements by using hypothetical sentences. PANKOW uses the IE phase to extract proper nouns and uses those nouns to generate these hypothetical sentences from an ontology. So if the system is working in the sports domain and it manages to extract the name "John Smith", a hypothetical sentence would be "John Smith is a player". This is then fed to a search engine and different similar instances are discovered. The phrase with the highest query result is used to annotate the text with the appropriate concept. This idea is based upon what they call the "disambiguation by maximal evidence" whereby the popularity of a phrase is an indication of its truth value. However, even though this does give a good indication, it is not infallible. One simple example is the well known case of the popular TV show "Who wants to be a millionaire?" (mentioned in Section 8.1) which shows that popular belief is not necessary true.

---

[4] Precision is a measure which quantifies the number of correct annotations out of all the annotations created by the system.

[5] Recall is a measure which quantifies the number of correct annotations out of all the annotations possible in a particular document or in a collection of documents.

7.5   Kim

[182] [181] describes the Knowledge and Information Management platform (KIM)
which is made up of an ontology, a knowledgebase, a semantic annotator, an in-
dexer and a retrieval server. It makes use of several third party packages including
SESAME RDF repository [42], Lucene search engine [105] and GATE [66].

The techniques used by KIM are various and include gazetteers, shallow analysis
of texts and also simple pattern matching grammars. By combining these techniques
together, the system manages to produce more annotations with a higher level of
accuracy. [138] report that KIM manages to achieve an average of 91.2% when
identifying dates, people, organisations, locations, numerical and financial values
in a corpus of business news. Even though this show extremely good results, the
type of data sought by the algorithm might effect drastically the performance of the
algorithm. The distinguishing feature of KIM is that it not only manages to find most
of the information available but it also labels all the occurrences of an instance using
the same URI. This will then be grounded to an ontology thus ensuring that there
are no ambiguities between instances. The instance is also saved to the database if
it is not present already. KIM will also check for variances so that "Mr J Smith"
and "John Smith" will be mapped to the same instance in the ontology and assigned
the same URI. This approach ensures consistency between the same or different
variants of the same tag.

7.6   P-Tag

P-TAG as defined in [50] is a large scale automatic system which generates person-
alised annotation tags. The distinguishing factor between the other systems in this
section is the personalisation aspect. This system does not rely on generic annotations based upon common usage such as identifying named entities, time, financial
values, etc. P-Tag adds to this a dose of personalisation. To do this, it takes three approaches; the first is based on keywords, the second one on documents and the third
one is a hybrid approach. In practice, the approaches are quite simple and similar to
each other.

The keyword approach first collects documents retrieved from a search engine,
it extracts its keywords and compare them to keywords extracted from the person's
desktop. The document approach is similar, but rather than comparing with keywords spread around the user's desktop, it compares them with keywords found
in specific documents. So the scope effectively changes, the former tries to match
a user profile which is specific to a user but covers a generic topic (based upon the
user's interests). The latter matches a user profile which is both specific to a user and
specific to a topic (since it is bound to a particular document). The hybrid approach
marries the best of both words together.

The positive aspect of this system is that the user does not needs to specify some
seed elements or direct the engine. This information is harvested directly from the

user's computer by gathering all the implicit information that exists. With the advent of the social web, these systems are gaining even more importance and we've seen other similar systems emerge such as [107].

7.7 Conclusion

Various approaches have been highlighted throughout the chapter. It is interesting to notice the progression from systems which were seeded, thus required explicit examples to systems which require no examples but gather the information implicitly from the users. The next wave of such systems seems targeted toward exploiting the social networks, by analysing the social networks and through them understand better what the user likes and dislikes. An important topic in this section was the redundancy of the web. In fact, the next chapter will delve further into it.

Part III
Peeking at the Future

"You can analyse the past,
but you have to design the future!"

Edward de Bono

Chapter 8
Exploiting the Redundancy of the Web

A key feature of the web is the redundancy of information and this is only possible due to the presence of the same or similar information in different locations and in different superficial forms. This feature was extremely useful to annotate documents since it allowed harvesting algorithms to gather information from various sources, check its reliability and use it for annotation purposes.

A clear example of this form of redundancy are the reports found in different newspapers. Most of them relate to the same or similar information but from a slightly different perspective. In fact, a quick look at Google news[1] reveals that stories are repeated several times, some of them even appearing thousands of times on different newspapers. An example of this is a story about 48 animal species in Hawaii. According to Google News, this story was published around 244 times in online newspapers and blogs around the web. The following is a sample of titles found on these pages which include:

- 48 Hawaii-only species given endangered listing[2]

- 48 Hawaiian Species Finally Added to Endangered List[3]

- Hawaiian birds among 48 new species listed as endangered[4]

- 48 Species On Kauai To Receive Protection[5]

- 48 Kauai species join endangered list[6]

An analysis of these few examples reveal that the content of the various articles are the same, even though they are published on different web sites and written in a slightly different form. It is interesting to note that if we seek information about this new list of 48 animal species, Google returns more than 1 million documents.

---

[1] http://news.google.com
[2] The Associated Press - Audrey McAvoy
[3] Greenfudge.org - Jim Denny, Heidi Marshall
[4] Los Angeles Times
[5] KITV Honolulu
[6] Honolulu Star-Bulletin

A. Dingli: Knowledge Annotation: Making Implicit Knowledge Explicit, ISRL 16, pp. 81–87.
springerlink.com   © Springer-Verlag Berlin Heidelberg 2011

What's impressive to note is that this news item spread around the globe in less than 24 hours. Some of it copied by human reporters working in various newspapers or by environmentalists in their blogs, the majority of the other copies was performed by automated agents whose job is to harvest news items and post them on some other website. The redundancy of the web is an interesting property because when different sources of information are present in multiple occurrences, they can be used to bootstrap recognisers which are capable of retrieving further information as was shown in Chapter 7.

8.1 Quality of Information

The redundancy aspect of the web can also serve as a form of quality control. Physical publications are normally scrutinised by editors, reviewers, etc before being published, thus providing them with an acceptable level of quality. However this is very different from the existent situation on the web.

In the past, the creation of a web page was restricted to people capable of understanding HTML. This meant that they could publish whatever they liked irrespective of its quality. However, since the people knowledgeable about HTML was limited, this meant that the amount of dubious material was limited as well. With the advent of Web 2.0, this changed forever. Blogs which are available on most sites allow people to freely express their views without restrictions. Users do not need to learn any HTML to contribute since the applications are engineered to promote user contributions. Considering that anyone can post anything online, the correctness of the information posted on certain websites is untrustworthy. Sites such as Amazon even go a step further by providing systems such as the Digital Text Platform[7] (DTP). By using the DTP, an author can publish his own book in seconds. The book is then available for purchase through the main Amazon site and it can be downloaded on any machine running Amazon's proprietary software. Because of this, we need a system capable of assessing the reliability of online information. This is why the redundancy property of the web is so important because a system can easily use the distribution of facts across different sources to measure their reliability.

This idea finds its roots in [213] which states that knowledge created by a group of people is most likely more reliable than that created by a single member. The concept only holds for groups of people who, according to him, follow four key criteria:

- the members of the group should be diverse from each other to ensure enough variance in the process

- each member should form his/her decision independently of the others without any influence whatsoever

---

[7] https://dtp.amazon.com/

- members should be capable of taking decisions based on local knowledge and specialise on that

- there should be a procedure to aggregate each member's decision into a collective one

The web satisfies all of these criteria. Web users reside in all the corners of the globe thus providing diversity. Most of them have no connection with one another which ensures independent judgements. Local knowledge is readily available in today's society where media consumption is at its peak. Finally, the aggregating procedures can be found online on the web. In the document, Surowiecki collates various examples in support of this idea. An interesting example can also be seen in the popular TV quiz show 'Who Wants to be a Millionaire?' where contestants are asked a number of multiple-choice questions with an ascending level of difficulty. When the contestant is unsure or doesn't know the answer, he can ask the audience for a suggestion. It was noticed that the answer given by the audience was surprisingly correct in over ninety per cent of the cases. [177] studied this phenomena and came up with the following conclusion. If we assume that a question was asked to a group of 100 people where about one-tenth of the participants knew the answer and two-tenths of the participants possessed only partial information, statistically, the correct answer would prevail. The reasoning behind this is very simple, if we just take into consideration the one-tenth that know the answer and we assume that the rest will select an answer randomly out of the 4 possible answers (ignoring the fact that two-tenths of them posses partial information), we get the following results:

10 correct answers + ((100-10) / 4) correct answers = 33 correct answers

This is higher that the probability of having all of them select a random result which would equal to just 25 correct answers. If we add those people who posses partial information to the equation, this would easily go up to 40 correct answers out of 100. An explanation for this can be found in [212] based on the Condercet Jury Theorem which states that

> if an average group member has better than a 50 percent chance of knowing the right answer and the answer is tabulated using majority rule, the probability of a correct answer rises toward 100 percent as the group size increases.

With this explanation in hand, it becomes clear that this idea is well suited for the web. The content found online was created by people who satisfied the four criteria mentioned earlier. Thus, the online information is not random but made up of an aggregation of partial truths. Because of this, if we manage to harvest this information and analyse the prevailing topics, we can manage to sift between what is dubious and what is real.

8.2 Quantity of Information

The redundancy property of the web would not be effective without the huge number of people contributing large amounts of information daily. The idea of using this collective intelligence is not new and in fact, it has been studied in [26] and [32]. In a typical system which simulates swarm intelligence, several elements coordinate between themselves and integrate with their environment in a decentralised structure. This methodology eventually leads to the emergence of intelligent behaviour as mentioned in [133]. Emergence can be linked to the distributed nature of the WWW since the web is based on links derived from a set of independent web pages with no central organisation in control. The result of this is that the linkage structure exhibit emergent properties. These properties can be seen in [151] and [60]. In fact, they claim that a correlation exists between the number of times an answer appears in a corpus and the performance of a system (when asked to solve the question pertaining to that answer). This means that if the answer to a question appeared several times, systems tend to perform better. This result might sound rather obvious, however, it helps us understand better this property.

The downside of such an approach is obviously noise. If we increase our training set so that it contains more potential answers with the hope of improving the results of our algorithms, we might introduce noise. In so doing, the new data might have a negative effect on the results. However, experiments by [24] have shown that noisy training data did not have such a negative effect on the results as one would expect. In fact, the effect was almost negligible.

8.3   The Temporal Property of Information

Information is also more sensitive than we think. A piece of information does not only have a truth value but its truth value has a temporal dimension too. If we use a common search engine and we enquire about the President of the United States, President Obama features in 50 million documents whereas President Bush features in 5 million documents (and this not withstanding that the United states had two Presidents whose surname was Bush in the last two decades). This shows that about 5 million documents are still refering to Mr Bush as though he is still the president. However, we can also derive an interesting observation which we noticed earlier as well, the fact that new information gets copied quickly around the web. In fact, the number of web pages referring to President Obama grossly outnumber those referring to President Bush even though the former is going through his first term as President of the United States. This seems to suggest that new information gets more prominence online.

8.4   Testing for Redundancy

To test the significance of redundant information, a simple experiment was conducted. Wikipedia essentially contains two types of articles, featured articles and non-featured. The featured articles are those which have been rated by the Wikipedia editors as being the best articles on the site. On average, the ratio between featured articles and non-featured is about 1: 1120. Since featured articles contain reliable information (according to the editors), the redundancy of the web can be used to sift between these articles and the non-featured ones. The experiment was conducted as follows:

1. A sample of about 200 random documents was harvested from Wikipedia (See Appendix A). 100 from the featured list and another 100 from the non-featured list.

2. [66] was used to extract just the textual content (and eliminating menus, etc).

3. The most relevant sentences were then chosen based upon [207].

4. These sentences were then used to query a search engine and retrieve related documents.

5. Finally, one of the similarity measures implemented in SimMetrics[8] was used to check similarity between the sentences in the Wikipedia article and the ones in the retrieved article. In principle, several similarity algorithms were tested such as the Levenshtein distance, Cosine similarity and the Q-gram. However, the Q-gram distance gave the best results.

6. When the similarity score is obtain for each sentence, an average score is then calculated for each document by averaging the scores for each sentence contained in the document.

The results obtained from this experiments were very significant. On average, a featured article obtained a similarity score of 67% when compared with other documents available online. This contrasted greatly with the non-featured articles which only managed to obtain an average score of 47%, a difference of around 30%. When examining these results further, another interesting correlation surfaces. The featured articles have a consistently higher number of edits. In fact, on average the top 10 featured articles (according to the similarity score) were edited about 5,200 times. In contrast, the top 10 non-featured articles were only edited around 700 times. A similar correlation was found between the number of references in a document and the document's quality. On average, the top 10 featured articles contained about 140 references each whereas the non-featured articles contained just 13 references each. This clearly show that there exists an implicit relationship between a document and the redundant information lying in different locations around the web.

8.5 Issues When Extracting Redundant Data

Even though different copies of the same piece of information can exist all over the web, we are still faced with various issues when it comes to harvesting that data. The type of information found online is normally present in different formats. This includes documents (Word Documents, Adobe PDF, etc), repositories (such as databases or digital libraries) and software agents capable of integrating information from different sources and providing a coherent view on the fly. Software programs

---

[8] An open source library of string similarity metrics developed at the University of Sheffield.

can harvest documents and access them quite easily when they are dealing with open formats such as XML. However, when dealing with proprietary formats such as Microsoft Word, the task complicates itself since the format of the document is not freely available and it will be hard to extract the information from it.

Another problem of these documents is the nature of the information represented in the document. In the past, before computers were used for all sorts of applications, text type was not an issue because the only kind of text available was free text. Free text contains no formatting or any other information. It is normally made up of grammatical sentences and examples of free texts can range from news articles to fictional stories. Text normally contains two types of features, syntactic and semantic features. Syntactic features can be extracted from the document using tools like part-of-speech taggers [40], chunkers [205], parsers [46], etc. Semantic information can be extracted by using semantic classifiers [67], named entity recognisers [160], etc.

In the 60's when computers were being used in businesses[9], and information was being stored in databases, structured data became very common. This is similar to free text but the logical and formatting layout of the information is predefined according to some template. This kind of text is quite limiting in itself. The layout used is only understandable by the machine for which it was created (or other compatible ones). Other machines are not capable of making sense of it unless there is some sort of translator program. Syntactic tools are not very good at handling such texts. The reason being that most tools are trained on free texts. Apart from this, the basic structures of free text (such as sentences, phrases etc.) do not necessary exist in structured text. Structured text mainly has an entity of information as its atomic element. The entity has some sort of semantic meaning which is defined based upon its position in the structure. Humans on the other hand show an unprecedented skill of inducing the meaning most of the time. Yet, they still prefer to write using free text. Therefore, these two types co-existed in parallel for many years.

With the creation of the WWW, another text type gained popularity, semi-structured text. Unfortunately it did so for the wrong reasons. Before the semi-structured text era, information and layout were generally two distinct objects. In order to enable easy creation of rich document content on the internet, a new content description language was created called HTML. The layout features which until recently were hidden by the applications became accessible to the users. The new document contained both layout and content in the same layer. This provided users the ability to insert new structure elements such as tables, lists, etc inside their documents together with free text. Now users could create documents having all the flexibility of free text with the possibility of using structures to clarify complex concepts. Obviously this makes it more difficult for automated systems to process such documents. Linguistic tools work well on the free text part but produced unreliable results whenever they reached the structured part. Standard templates do not exist for the structured part because the number of possible combinations of the different structures is practically infinite.

---

[9] http://www.hp9825.com/html/hp_2116.html

To solve this problem, artificial intelligence techniques are normally adopted. When dealing with free text or semi-structured texts, natural language processing techniques (Such as Amilcare [58], BWI [97], etc) manage to extract most of the information available. Databases on the other hand might provide an interface through which the data can be queried. But when all of these approaches fail, screen scrapers[10] can be used to extract such information.

Another issue associated with redundant data is the enormous quantities of documents available online and how to process them. Luckily, modern search engines already tackled this issue, in fact they are capable of indexing billions of documents. The only problem is that systems which rely on search engines to identify redundant data have to be careful because the scoring algorithm of every search engine is a well guarded secret and frequent tweaks (with the hope of improving the results) might make the results unpredictable over time. Thus, systems have two options, either use the unpredictable engines or create their own.

The former goes against well established principles since the system cannot provide the users with consistent results (considering it is at the mercy of the changes to the search engine). The latter is unreachable for most institutions since it requires a lot of resources and efforts. In reality, even though search engines suffer from these problems, they are still used for this kind of research. However, an analysis conducted by [109] shows that around 25% of the visible web (and we're not even considering the deep web[11]) is not being indexed by search engines. In fact, according to this study, Google indexes 76%, Msn 62% and Yahoo! 69%. The study also estimates that the amount of redundancy[12] between the various search engines amounts to about 29% or 2.7 billion pages. Thus, almost one out of every 3 pages can be found on all the search engines. These results are rather interesting because they show us that almost one-fourth of the web is not being indexed, another one-fourth is being indexed by all of the major search engine and the rest is dispersed amongst them. This means that to harness the power of search engines in order to find redundant information, researchers have to use a combination of the major search engines to obtain the best results. This is also congruent with the technique described in [88] where they successfully used up to twelve search engines in their system.

8.6 Conclusion

This chapter investigated an important property of the web generally referred to as the redundancy of information, explaining what it is and why it is so important for the annotation process. It identified common pitfalls but it also highlighted ways in which this property can be exploited. The chapter also showed that if this property is harnessed, it can produce some amazing results. The final Chapter, will take a peek towards the future of annotation.

---

[10] A software program capable of extracting information from human readable output.

[11] Part of the web which cannot be indexed using traditional search engine technologies such as dynamically generated sites.

[12] Redundancy in the context of search engines refers to the intersection of search engine's indexes.

Chapter 9
The Future of Annotations

This document has shown how vital the whole annotation process is for the web. Unfortunately, even with the various techniques mentioned throughout the text, the annotation task is far from being trivial since it is still tedious, error prone and difficult when performed by humans. If we also consider the fact that the number of web pages on the internet is increasing drastically every second, manual annotation becomes largely unfeasible. One could consider asking the authors of the various web sites to include semantic annotations inside their own documents but that would be similar to asking users to index their own pages! How many users would go about doing so and what sort of annotations should they insert?

Some years ago, a drive was made towards using meta tags inside HTML documents. These tags insert information into the header of web pages, they are not visible by users and are used to communicate information (such as the "character set" to use, etc) to crawlers and browsers. These tags laid the initial steps towards inserting semantic information[1] into a web page. Since the users could not really feel the added benefit they get from inserting these tags and considering search engines were not giving them particular weight, the meta tag suffered a quiet death.

Another reason why the web reached such a massive popularity is mainly due to sociological factors. Originally, the web was created by people for people and since people are social animals, they have an innate desire to socialise, take part in a community and make themselves known in that community. In fact, the history of the web can be traced back to a network of academics. These people all came from the same community. They needed to get to know other people working in their area, share their knowledge and experiences, and also collaborate together. Snail Mail[2] was the only form of communication available between these academics but it was

---

[1] The meta keywords tag allows the user to provide additional text for crawler-based search engines in order to index the keywords along with the body of the document. Unfortunately, for major search engines, it does not help at all since most crawlers ignore the tag.

[2] A slang term, used to refer to traditional or surface mail sent through postal services. Nicknamed snail mail because the delivery time of a posted letter is slow when compared to the fast delivery of e-mail.

A. Dingli: Knowledge Annotation: Making Implicit Knowledge Explicit, ISRL 16, pp. 89–95.
springerlink.com © Springer-Verlag Berlin Heidelberg 2011

too slow. E-mail changed all this and made it possible for academics working far away to exchange information in a few seconds. They saw a potential in this technology and therefore decided to grant access to this service even to their students. The latter saw even greater potential in this network and started experimenting with new ideas such as Bulletin Board Services (BBS) and online games (such as Multi User Dungeons (MUD)). The original scope of the net changed completely. It was not only limited to collaboration between individuals for work related purpose but also became a form of entertainment. This new and exciting technology could not be concealed for long within the university walls and quickly, people from outside these communities started using it too. It was a quick way of communicating with people, playing games, sharing multimedia files, etc. This web grew mainly because, it gained popularity in a community which was very influential and which gave it a very strong basis from where to expand. Subsequently, since this technology is easy to use by anyone and requires no special skills, its usage expanded at a quick rate.

Even though all this happened before the web as we know it today was even conceived, the current web grew exactly in the same way. With the appearance of web browsers, people who have been using the internet realised that they could move away from the dull world of textual interfaces and express themselves using pages containing rich multimedia content. People soon realised that it was not just a way of presenting their work, this new media gave them an opportunity even to present themselves to others. It allowed anyone to have his/her own corner on the internet accessible to everyone else. A sort of online showcase which was always there representing the personal views, ideas, dreams, etc twenty four hours a day, seven days a week! The technology to create these pages was a little bit difficult to grasp initially but soon became extremely user friendly having editors very similar to text processors (a technology which has been around for decades and which was known by everyone who was computer literate) and easy Web 2.0 applications. To cut a long story short, everybody wanted to be present in this online world even though most people did not even know why they wanted to be there!

The initial popularity grew mainly due to word of mouth, but that growth is insignificant when compared with the current expansion which the web is experiencing. Search engines were quite vital for this success. Before search engines were conceived, one found information only by asking other people or communities (which were specialised in the area) regarding sites where information could be found. After, the process of looking for information was just a matter of using a search engine. Therefore, information did not have to be advertised anywhere since automated crawlers were capable of searching the web for information and keeping large indexes with details of where that information is found.

The web of the future is currently being constructed as an extension of our existent web. In the same way as search engines are vital for the web of today, semantic search engines (such as Haikia[3], SenseBot[4], etc) will be vital for

---

[3] http://www.hakia.com

[4] http://www.sensebot.net

the web of tomorrow. These search engines will allow searches to be conducted using semantics rather than using the bag of words approach (popular in today's search engines). For this to happen, we must have programs similar to crawlers that create semantic annotations referencing documents, rather than just keywords. To discover these annotations, we further require some automatic semantic annotators which semantically annotate the documents. But this is just the tip of iceberg ...

9.1 Exploiting the Redundancy of the Web

The automated methodologies proposed in this document all make use of the redundancy of information. Information is extracted from different sources (databases, digital libraries, documents, etc.), therefore the classical problem of integrating information arises. Information can be represented in different ways and in different sources from both a syntactic and a semantic point of view. Syntactic variations of simple types are generally dealt with quite easily e.g. the classical problem of recognising film titles as "The big chill" and "Big chill, the" can be addressed. More complex tasks require some thought. Imagine the name of a person, it can be cited in different ways: in fact J. Smith, John Smith and John William Smith are potential variation of the same name. But do they identify the same person as well? John Smith is found in no less than 4 million documents. When large quantities of information is available (e.g. authors names in Google Scholar) this becomes an important issue [14]. This problem intersects with that of intra- and inter-document co-reference resolution well known in Natural Language Processing (NLP).

By seeking more websites related to the task, simple heuristics can be applied to tackle these problems in a satisfying way. For example the probability that J. Smith, John Smith and John William Smith are not the same person in a specific website is very low and therefore it is possible to hypothesise co-reference. Different is the case of ambiguity in external resources (e.g. in a digital library). Here the problem is more pervasive. When querying with very common names like the above example, the results are quite disappointing since papers by different people (having the same name) are mixed up. We have to keep in mind that we live in a fast changing world full of discrepancies and to deal with these changes, small discrepancies are normally accepted by everyone. If we image a task where we need to find information about the number of people living in a particular country, it is very likely that different reliable sources will have different values, therefore creating discrepancies. These values might be correct if they present a number which is approximately equal to a common average, accepted by everyone (but which is not necessarily the exact number!). Thus, our precise systems have to deal with discrepancy and accept the fact that some might not be exact but its the best correct answer we can get. In [213] we find that this is what most people do in most cases and the results are generally incredibly accurate.

9.2   Using the Cloud

Most of the systems described in Chapter 7 serve their intended purpose quite well
however to deal with the overwhelming size of the web, new approaches need to
be considered. In any computerised system, resources (such as processing power,
secondary storage, main memory, etc.) are always scarce. No matter how much we
have available, a system could always make use of more. These systems are very
demanding, much more than normal ones since they need to process huge amounts
of texts using complex linguistic tools in the least possible time. Physical resources
are not the only bottleneck in the whole system. The raw material which the systems
utilise are web pages, downloaded from the internet. Unfortunately, this process is
still extremely slow, especially when several pages are requested simultaneously
from the same server. In synthesis, we can conclude that such systems are only
usable by top of the range computers having a high bandwidth connection to the
internet. A possible solution is to exploit new technologies in distributed computing
such as cloud[5] computing.

The benefits of such an approach are various. Since the system is utilising resources elsewhere on the internet, there is no need of a powerful machine with huge
network bandwidth. The client is just a thin system whereby the user defines the
seed data and simply initialises the system which is then executed somewhere remotely in the cloud. However, such a system has to be smart enough to deal with a
number of issues;

Robustness - In a system which makes heavy use of the web, it is very common to
   try to access links or resources that are no longer available. There can be various
   reasons for this, a server could be down, a page does not exist any longer, etc.
   In this case, if an external resource is not available, alternative strategies can be
   adopted. The most intelligent ones would look for other online resources as a
   substitute. Another possible scenario would be to contact the client and ask for
   the link to another resource.

Accountability - When new data is produced, meta-information such as creation
   date, etc should be included with the data. The system should be able to adapt to
   the users needs by analysing its past actions. Therefore, if an item of data was
   reviewed by a user and rejected, the system should reevaluate the reliability rating
   of the source from where the data was obtained. The global reliability rating of a
   site must be a reflection of the data it produces.

Quality control - The system should also be capable of performing automatic checks
   without the user's intervention and deciding what appropriate action to take when
   necessary. Imagine some seed data is used to train a learning algorithm. If the
   precision of the algorithm is low when tested on the training data itself, then the
   system should realise that the learning algorithm is not good enough and exclude
   it from the whole process automatically. This is just one of the many automatic
   tests which can be performed and the degree of automation of these tests depends
   entirely upon the application and its complexity.

---

[5] A cloud is a framework used to access and manage services distributed over the internet.

9.3 The Semantic Annotation Engine

A Semantic Annotation Engine is a engine that given a document, annotates it with semantic tags. The need for such an engine arises from the fact that the web currently contains millions of documents and its size is constantly increasing [61][62]. When we are faced with such huge tasks, it is inconceivable that humans will manage to annotate such a huge amount of documents. An alternative would be to use automated tools as discussed in Chapter 7 but this is only a partial solution. There are a number of problems to face with this approach. First of all the system is specific to a particular domain and cannot just semantically annotate any document (especially if that document is not related to the current domain). If we assume that the system can annotate any document, a problem arises regarding which annotation is required for which document. In theory, a document could have an infinite number of annotations because different users would need different views of a particular document. Therefore, deciding which document contains which annotations should not be a task assigned to the annotation system but rather to the user requesting the annotations. Apart from this, there are still some open issues with regards to annotations, like should annotations be inserted as soon as a document is created or not? Some annotations can become out of date very quickly like weather reports, stock exchange rates, positions inside a company, etc. What would happen when this volatile information changes? Would we have to re-annotate the documents again? But that would mean that we would not need annotation engines but re-annotating engines constantly maintaining annotations. The order of magnitude of the problem would therefore increase exponentially. The second problem is that probably, the rate at which the web is growing is much faster than such a system could annotate.

Another more realistic and generic solution would be to provide annotations on demand, instead of pre-annotating all the documents i.e. the Semantic Annotation Engine (SAE). This means that annotations will always be the most recent and there would not be any legacy with the past. If a document is out of date or updated with more recent information, the annotations would still be the most recent.

In traditional Information Retrieval (IR) engines (like Google[6], Yahoo[7], etc), a collection of documents (such as the web) is crawled and indexed. Queries using a bag of words are used to retrieve the page within the collection that contains most (if not all) of the words in the query. These engines do a pretty good job at indexing a substantial part of the web and retrieving relevant documents, but they are still far from providing the user exactly what he needs. Most of the time, the users must filter the results returned by the search engine. The SAE would not affect in any way the current IR setup. The search through all the documents would still be performed using the traditional IR methods, the only difference would be when the engine returns the results to the user. Instead of passing the results back to the user, they are passed to a SAE. The system can also contain an index of the different SAEs which are available online. The SAEs are indexed using keywords similarly to how

---

[6] http://www.google.com

[7] http://www.yahoo.com

normal indexing of web documents works. Whenever an IR engine receives a query, it not only retrieves the best document with the highest relevance but also the SAEs with the highest relevance. These are then used to index the documents retrieved by the search engine before passing them back (annotated) either to the user or to an intermediate system that performs further processing. SAE will provide on the fly annotations for web documents therefore avoiding the need of annotating the millions of documents on the web beforehand.

One relevant question for the effective usability of this methodology in real applications concerns the required level of accuracy (as a balance of precision and recall) the system has to provide. As Web applications are concerned, it is well known that high accuracy is not always required. Search engines are used every day by millions of people, even if their accuracy is far from ideal: further navigation is often required to find satisfying results, large portions of the Web are not indexed (the so called dark and invisible Webs), etc. Services like Google Scholar, although incomplete, are very successful. What really seems to matter is the ability to both retrieve information dispersed on the Web and create a critical mass of relatively reliable information.

9.4 The Semantic Web Proxy

A Semantic Web Proxy (SWP) is similar to a normal web proxy[8], it provides all the functionality associated with such a program. The main difference is that it also provides some semantic functions hidden from the user.

To understand what these semantic functions are, lets take a look at a typical example. Imagine a user who would like to purchase a computer having a 17 inch flat screen monitor, 1 GBytes of RAM, a 3 GHz processor, etc. The user would go to an online store, look for a search form where he can write the search criteria and perform the search for the desired product. This operation must be repeated for each and every online store he would like to query.

A SWP would simplify this process in several ways. First of all, it would keep a record of all the forms being filled in a single session[9] and the content of those forms. If the forms are semantically tagged or associated with an ontology, then the details inserted by the user are automatically tagged as well. Once ready, the user would then go to a different site and perform a similar search. The system would notice that the form in the new site is filled with the same details as those in the other site and it would take the following actions:

---

[8] A proxy is an intermediate server that sits between the client and the origin server. It accepts requests from clients, transmits those requests on to the origin server, and then returns the response from the origin server to the client. If several clients request the same content, the proxy can deliver that content from its cache, rather than requesting it from the origin server each time, thereby reducing response time.

[9] A session is a series of transactions or hits made by a single user. If there has been no activity for a period of time, followed by the resumption of activity by the same user, a new session is considered started.

1. tag the details inserted by the user with the semantic tags found in the new site.

2. create equivalence relationships between the semantic tags in the first site and the tags in the second site.

3. make these relationships available to everyone.

The effect of this process is the following. If another user performs a search for a similar item in one of the sites already in the SWP, the system uses the shared relationships (obtained before through the interaction with the other users) and automatically searches the other online stores. All the results are returned to the user for evaluation.

The system basically creates links between ontologies in a shared way collaboratively without the user realising it. It does so by examining the browsing habits of the user and deducing implicit relations when possible. Once again, the system adopts the same ideas used to create the web. Basically, it exploits the little work of many users (without adding any extra effort) to automatically create relationships between concepts over the web.

9.5 Conclusion

In this book, we have been on the annotation journey, one that started hundreds of years ago when annotation was still made up of scribbles. The importance of annotations transpires in every chapter and helps us understand why annotation is such a fundamental aspect of our day-to-day life on the web. Unfortunately annotation is not something trivial and we have seen ways of how documents can be manually, semi-automatically or automatically annotated using different techniques. There are still various open issues associated with annotations but hopefully in the coming years, new powerful techniques will be developed which will help us annotate the web thus creating a improved web experience for everyone.

Appendix A
Wikipedia Data

This appendix lists all the articles that were mention in section 8.4. These articles were selected randomly from Wikipedia. Table A.1 lists 100 featured articles whilst table A.2 lists 100 non-featured articles. The main difference between the two kind of articles is that whereas a non-featured article does not have a specific formatting to follow, a featured one has to follow well defined guidelines. These depend upon the type of articles however they normally include:

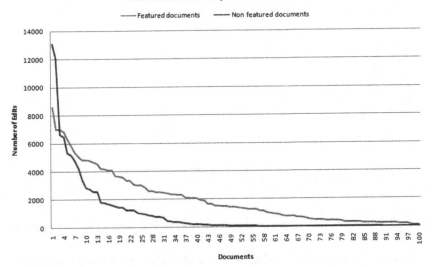

Fig. A.1 A document showing the number of edits done on each and every document

- chronology if the article is relating a particular set of events,
- cause and effect if a particular event is being examined taking into consideration what triggered the event and what was the result,
- classification which implies the grouping of certain elements in the article,
- question / answering in articles about interviews.

These two tables are divided into 3 columns. The first one is the article's name, the second is the number of edits per document and the third one is the number references per document. The edits per document can be seen in Figure A.1 and they clearly show that the edits of the featured documents is significantly higher. On average, a featured document gets edited around 2000 times whereas a non-featured one gets edited only about 1000 times. The references added to the document can be seen in Figure A.2 and this too shows that featured documents have a substantial number of references more than the non-featured ones. In fact the average references for a featured article is 90 whereas for the non-featured is 20, this is equivalent to 450% more.

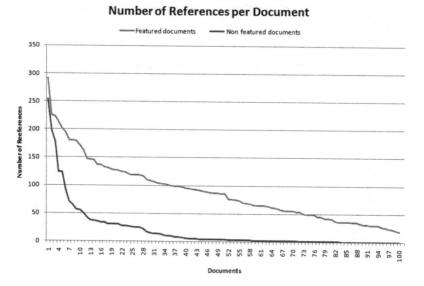

Fig. A.2 A document showing the number of references added to each and every document

Table A.3 and table A.4 too list the featured articles and the non-featured ones however they have three additional columns. These list the similarity scores obtained when using the Levinshtein Distance, the Cosine Similarity and the Q-Gram similarity measures. It is interesting to note that the three algorithms produce very similar results. Also the data in the table has been ordered based upon these similarity measures. The variance in the results between the Levenshtein Distance and the others is around 0.2% whilst between the Cosine and the Q-Gram similarity, the difference is negligible. When the average similarity result is compared, it transpires that the Q-gram similarity produces a similarity score in between the Levenshtein Distance and Cosine similarity as can be seen in Figure A.4 and Figure A.3. Thus because of this, the Q-gram similarity will be used in the tests.

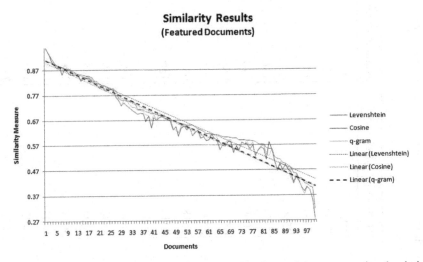

Fig. A.3 A summary of the similarity scores obtained for featured documents using the similarity algorithms together with their respective linear trend line

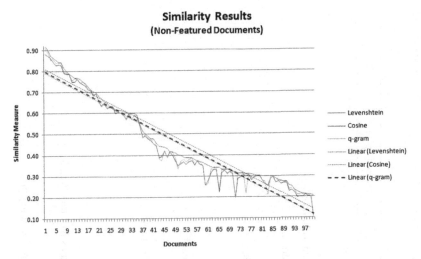

Fig. A.4 A summary of the similarity scores obtained for non-featured documents using the similarity algorithms together with their respective linear trend line

Table A.1 Featured Articles harvested randomly from Wikipedia together with the number of edits and references

Title	Number of edits	Number of references
Triceratops	1649	65
Boston	7033	179
Pyotr Ilyich Tchaikovsky	5831	147
Rachel Carson	2590	92
R.E.M.	3356	136
Titan (moon)	2069	119
Macintosh	6832	98
Lion	5365	201
Omaha Beach	1518	93
Primate	2312	136
Matthew Boulton	795	124
Tyrannosaurus	4840	119
Microsoft	8596	119
Pluto	7024	145
United States Military Academy	3019	225
Kolkata	4807	105
M249 light machine gun	1495	49
Beagle	3038	76

Nauru	1454	50
Diamond	4800	94
Texas Tech University	2496	195
United States Marine Corps	6313	104
"University of California and Riverside"	4187	128
George III of the United Kingdom	3344	120
Venus	5043	132
Warwick Castle	1293	61
Edward Wright (mathematician)	326	56
Hungarian Revolution of 1956	2839	180
Sheffield Wednesday F.C.	3679	42
F-4 Phantom II	2459	111
Harvey Milk	2615	171
Blue whale	2443	65
George B. McClellan	1452	96
Hydrochloric acid	2079	19
Learned Hand	1495	223
William Gibson	1642	146
Han Dynasty	4095	292
Battle of the Alamo	4162	163
Battle of Vimy Ridge	3009	123

Stanley Cup	2354	57
Tim Duncan	4086	95
Bald Eagle	3570	54
USS Constitution	1901	213
Medal of Honor	2300	68
Battle of Dien Bien Phu	1505	90
Casablanca (film)	2529	117
Somerset	1395	127
Cryptography	2102	42
Black Moshannon State Park	512	54
Huntington's disease	4533	103
USS Congress (1799)	407	99
Blaise Pascal	3608	33
Cyclura nubila	437	41
Reactive attachment disorder	2118	99
W. S. Gilbert	1297	100
Red-tailed Black Cockatoo	480	69
Chicago Board of Trade Building	764	64
Angkor Wat	1379	49
Hurricane Kenna	260	21
Bart King	244	26

Definition of planet	2341	89
The Philadelphia Inquirer	480	37
Black Francis	1273	86
General aviation in the United Kingdom	647	126
William D. Boyce	770	86
"Albert Bridge and London"	187	36
Katyn massacre	1937	109
Loihi Seamount	977	35
Edward Low	690	31
Atmosphere of Jupiter	889	107
Drapier's Letters	286	102
Bruce Castle	301	45
Holkham Hall	1041	27
Aliso Creek (Orange County)	1275	75
History of the Montreal Canadiens	449	180
Sunday Times Golden Globe Race	320	76
Princess Helena of the United Kingdom	269	87
George H. D. Gossip	1137	131
Battle of Goliad	345	24
John Brooke-Little	520	30
Shrine of Remembrance	436	70

Mount St. Helens	4675	56
Apollo 8	1138	59
Columbia Slough	586	87
Vauxhall Bridge	201	30
Aggie Bonfire	929	62
Larrys Creek	444	66
Aiphanes	328	49
Cleveland Bay	206	36
Falaise pocket	686	88
Frank Hubert McNamara	92	45
Battle of Cape Esperance	289	36
Wulfhere of Mercia	233	65
Tunnel Railway	140	31
Rhodotus	79	56
Golden White-eye	244	23
Kaiser class battleship	226	36
Armament of the Iowa class battleship	229	35
The Battle of Alexander at Issus	840	74
New York State Route 174	327	30

Table A.2 Non-featured Articles harvested randomly from Wikipedia together with the number of edits and references

Title	Number of edits	Number of references
Albert einstein	13061	123
Roberto Baggio	12089	254
Armenian Genocide	6648	198
Macroelectronics	5317	36
Bible	6466	34
Russia	2835	31
Toyota	5155	94
Quantum tunnelling	1778	37
Sven Goran Eriksson	4795	123
Jack Purvis	1218	31
XML	2772	31
Air france	2553	57
Dr Manmohan Singh Scholarship	1435	56
Tour de france	3438	177
University of london	823	31
South Carolina Highway 31	1669	25
Tolerance Monument	975	38
Ricardo Giusti	955	2

Suburb	2571	34
Received Channel Power Indicator	1027	8
Tesco	411	22
World Youth Day 2008	735	71
Telugu language	724	6
Gone with the wind	370	14
Estadio do Maracana	857	3
Sgt pepper	1616	17
Eddie Bentz	1450	8
Espionage	695	10
Glass engraving	1793	15
Phosducin family	443	2
Woodstock Festival	4195	51
Transmitter	354	3
Vaccine	1526	43
Astronomical naming conventions	332	1
Jane Saville	1241	27
Dan Sartain	1219	66
Defence Institute of Advanced Technology	292	3
Viaduct	206	4
William Remington	213	25

Amalgam (chemistry)	37	1
Polish 7th Air Escadrille	20	1
Krutika Desai Khan	135	28
Drexel University	6	1
Revolver	63	2
Macchi M.19	45	13
HMS Amphitrite (1898)	72	1
Bernhard Horwitz	48	2
Wisborough Green	34	3
Perfect fluid	57	0
Chris Horton	24	5
Longcross railway station	8	2
Human Rights Watch	75	28
Cromwell Manor	47	10
Messier 82	20	2
The Alan Parsons Project	56	4
Lump of labour fallacy	12	0
Sepia confusa	196	26
Long Island (Massachusetts)	121	4
Rodger Winn	8	2
Reptile	55	0

Blas Galindo	12	2
Apollodorus of Seleucia	10	5
Gordon Banks	31	7
Harold S. Williams	21	1
Darjeeling	32	0
Bristol derby	49	4
Domino Day	59	5
Makombo massacre	24	2
Palmetto Health Richland	27	1
Islands of Refreshment	67	4
Victor Merino	70	0
Gajaba Regiment	37	0
2B1 Oka	8	0
Melon de Bourgogne	52	1
Atonement (substitutionary view)	228	14
Cathedral of St. Vincent de Paul	29	1
Canton Ticino	27	0
Mathematics	167	11
Mountain Horned Dragon	31	0
Ultra Records	140	3
Guajira Peninsula	28	0

Harold Goodwin	32	1
Arthur Charlett	9	4
Hoyo de Monterrey	158	0
Juhapura	68	0
Fifa confederations cup	44	3
Yoketron	23	1
Bill Brockwell	18	0
John Seigenthaler	47	2
Operating model	54	4
Municipal district	22	0
David Grierson	16	0
Canberra Deep Space Communication Complex	92	3
Robert Rosenthal (USAF officer)	26	4
Johann Cruyff	106	0
Peada of Mercia	3	2
Hudson Line (Metro–North)	27	2
Last Call Cleveland	41	0
Blyth Inc	40	1
Chatuchak Park	106	2

Table A.3 Featured Articles together with their similarity scored when compared to articles obtained from a search engine

Title	Levenshtein	Cosine	Q-gram
Triceratops	0.96	0.96	0.96
Boston	0.94	0.94	0.94
Pyotr Ilyich Tchaikovsky	0.92	0.91	0.92
Rachel Carson	0.90	0.89	0.91
R.E.M.	0.89	0.88	0.88
Titan (moon)	0.88	0.88	0.89
Macintosh	0.88	0.85	0.87
Lion	0.88	0.88	0.89
Omaha Beach	0.88	0.87	0.87
Primate	0.88	0.85	0.88
Matthew Boulton	0.87	0.85	0.86
Tyrannosaurus	0.86	0.85	0.86
Microsoft	0.86	0.85	0.85
Pluto	0.85	0.83	0.83
United States Military Academy	0.85	0.84	0.84
Kolkata	0.85	0.84	0.85
M249 light machine gun	0.85	0.84	0.84
Beagle	0.84	0.83	0.84

Nauru	0.83	0.82	0.82
Diamond	0.83	0.81	0.82
Texas Tech University	0.82	0.80	0.81
United States Marine Corps	0.81	0.81	0.81
"University of California and Riverside"	0.81	0.79	0.80
George III of the United Kingdom	0.80	0.79	0.80
Venus	0.80	0.80	0.79
Warwick Castle	0.80	0.79	0.80
Edward Wright (mathematician)	0.79	0.76	0.76
Hungarian Revolution of 1956	0.76	0.75	0.75
Sheffield Wednesday F.C.	0.75	0.73	0.73
F-4 Phantom II	0.75	0.74	0.75
Harvey Milk	0.75	0.73	0.75
Blue whale	0.75	0.72	0.74
George B. McClellan	0.74	0.71	0.72
Hydrochloric acid	0.74	0.70	0.72
Learned Hand	0.72	0.70	0.71
William Gibson	0.72	0.70	0.71
Han Dynasty	0.71	0.70	0.71
Battle of the Alamo	0.70	0.67	0.69
Battle of Vimy Ridge	0.70	0.69	0.69

Stanley Cup	0.70	0.64	0.67
Tim Duncan	0.70	0.68	0.69
Bald Eagle	0.69	0.67	0.67
USS Constitution	0.69	0.68	0.70
Medal of Honor	0.69	0.69	0.69
Battle of Dien Bien Phu	0.69	0.68	0.69
Casablanca (film)	0.68	0.67	0.68
Somerset	0.68	0.67	0.68
Cryptography	0.67	0.63	0.64
Black Moshannon State Park	0.66	0.65	0.63
Huntington's disease	0.66	0.61	0.62
USS Congress (1799)	0.66	0.64	0.65
Blaise Pascal	0.66	0.65	0.65
Cyclura nubila	0.65	0.63	0.64
Reactive attachment disorder	0.65	0.64	0.64
W. S. Gilbert	0.65	0.61	0.61
Red-tailed Black Cockatoo	0.64	0.62	0.62
Chicago Board of Trade Building	0.64	0.63	0.63
Angkor Wat	0.63	0.60	0.61
Hurricane Kenna	0.62	0.60	0.61
Bart King	0.62	0.59	0.59

Definition of planet	0.62	0.59	0.60
The Philadelphia Inquirer	0.61	0.61	0.61
Black Francis	0.61	0.60	0.61
General aviation in the United Kingdom	0.61	0.59	0.60
William D. Boyce	0.61	0.55	0.58
"Albert Bridge and London"	0.60	0.57	0.58
Katyn massacre	0.60	0.56	0.58
Loihi Seamount	0.60	0.58	0.59
Edward Low	0.60	0.58	0.59
Atmosphere of Jupiter	0.60	0.54	0.55
Drapier's Letters	0.60	0.57	0.56
Bruce Castle	0.59	0.55	0.54
Holkham Hall	0.59	0.58	0.58
Aliso Creek (Orange County)	0.59	0.57	0.58
History of the Montreal Canadiens	0.59	0.57	0.58
Sunday Times Golden Globe Race	0.59	0.55	0.57
Princess Helena of the United Kingdom	0.59	0.57	0.58
George H. D. Gossip	0.58	0.53	0.52
Battle of Goliad	0.57	0.55	0.55
John Brooke-Little	0.57	0.56	0.57
Shrine of Remembrance	0.57	0.56	0.57

Mount St. Helens	0.56	0.51	0.53
Apollo 8	0.56	0.58	0.53
Columbia Slough	0.55	0.55	0.55
Vauxhall Bridge	0.52	0.51	0.51
Aggie Bonfire	0.51	0.47	0.47
Larrys Creek	0.51	0.50	0.50
Aiphanes	0.50	0.47	0.49
Cleveland Bay	0.50	0.49	0.49
Falaise pocket	0.49	0.47	0.48
Frank Hubert McNamara	0.48	0.46	0.44
Battle of Cape Esperance	0.46	0.42	0.42
Wulfhere of Mercia	0.46	0.44	0.45
Tunnel Railway	0.44	0.43	0.43
Rhodotus	0.44	0.40	0.40
Golden White-eye	0.42	0.39	0.37
Kaiser class battleship	0.42	0.40	0.40
Armament of the Iowa class battleship	0.40	0.39	0.39
The Battle of Alexander at Issus	0.40	0.35	0.35
New York State Route 174	0.32	0.28	0.27

Table A.4 Non-featured Articles together with their similarity scored when compared to articles obtained from a search engine

Title	Levenshtein	Cosine	Q-gram
Albert einstein	0.92	0.88	0.91
Roberto Baggio	0.90	0.87	0.89
Armenian Genocide	0.87	0.86	0.85
Macroelectronics	0.86	0.84	0.86
Bible	0.85	0.83	0.84
Russia	0.84	0.83	0.84
Toyota	0.84	0.83	0.84
Quantum tunnelling	0.82	0.79	0.80
Sven Goran Eriksson	0.80	0.79	0.79
Jack Purvis	0.79	0.79	0.79
XML	0.78	0.75	0.76
Air france	0.77	0.75	0.77
Dr Manmohan Singh Scholarship	0.77	0.76	0.76
Tour de france	0.76	0.75	0.74
University of london	0.75	0.74	0.75
South Carolina Highway 31	0.74	0.73	0.74
Tolerance Monument	0.73	0.71	0.72
Ricardo Giusti	0.71	0.71	0.71

Suburb	0.70	0.68	0.69
Received Channel Power Indicator	0.70	0.69	0.69
Tesco	0.68	0.65	0.65
World Youth Day 2008	0.67	0.64	0.63
Telugu language	0.65	0.65	0.65
Gone with the wind	0.65	0.62	0.62
Estadio do Maracana	0.63	0.63	0.63
Sgt pepper	0.63	0.63	0.63
Eddie Bentz	0.62	0.60	0.61
Espionage	0.61	0.60	0.60
Glass engraving	0.61	0.60	0.60
Phosducin family	0.61	0.59	0.59
Woodstock Festival	0.60	0.57	0.57
Transmitter	0.60	0.60	0.60
Vaccine	0.60	0.57	0.57
Astronomical naming conventions	0.60	0.59	0.59
Jane Saville	0.59	0.58	0.59
Dan Sartain	0.54	0.54	0.54
Defence Institute of Advanced Technology	0.51	0.48	0.48
Viaduct	0.50	0.49	0.50
William Remington	0.49	0.48	0.48

Amalgam (chemistry)	0.48	0.47	0.48
Polish 7th Air Escadrille	0.48	0.46	0.45
Krutika Desai Khan	0.46	0.44	0.45
Drexel University	0.44	0.39	0.38
Revolver	0.44	0.39	0.37
Macchi M.19	0.44	0.42	0.42
HMS Amphitrite (1898)	0.43	0.39	0.39
Bernhard Horwitz	0.42	0.41	0.42
Wisborough Green	0.41	0.40	0.41
Perfect fluid	0.41	0.37	0.39
Chris Horton	0.40	0.35	0.33
Longcross railway station	0.40	0.36	0.37
Human Rights Watch	0.39	0.37	0.38
Cromwell Manor	0.39	0.35	0.37
Messier 82	0.39	0.37	0.38
The Alan Parsons Project	0.39	0.37	0.39
Lump of labour fallacy	0.38	0.35	0.35
Sepia confusa	0.37	0.37	0.37
Long Island (Massachusetts)	0.37	0.35	0.37
Rodger Winn	0.36	0.35	0.35
Reptile	0.36	0.26	0.26

Blas Galindo	0.35	0.28	0.27
Apollodorus of Seleucia	0.34	0.32	0.33
Gordon Banks	0.34	0.33	0.34
Harold S. Williams	0.34	0.33	0.33
Darjeeling	0.34	0.22	0.23
Bristol derby	0.33	0.32	0.33
Domino Day	0.33	0.33	0.33
Makombo massacre	0.33	0.31	0.31
Palmetto Health Richland	0.33	0.32	0.32
Islands of Refreshment	0.32	0.31	0.32
Victor Merino	0.32	0.19	0.19
Gajaba Regiment	0.31	0.29	0.30
2B1 Oka	0.31	0.30	0.30
Melon de Bourgogne	0.31	0.31	0.31
Atonement (substitutionary view)	0.31	0.30	0.22
Cathedral of St. Vincent de Paul	0.31	0.30	0.31
Canton Ticino	0.31	0.28	0.29
Mathematics	0.31	0.29	0.29
Mountain Horned Dragon	0.30	0.30	0.30
Ultra Records	0.30	0.30	0.30
Guajira Peninsula	0.30	0.28	0.28

Harold Goodwin	0.30	0.27	0.28
Arthur Charlett	0.30	0.25	0.21
Hoyo de Monterrey	0.30	0.29	0.30
Juhapura	0.30	0.30	0.30
Fifa confederations cup	0.29	0.27	0.29
Yoketron	0.28	0.28	0.28
Bill Brockwell	0.28	0.26	0.28
John Seigenthaler	0.27	0.27	0.27
Operating model	0.27	0.27	0.27
Municipal district	0.25	0.23	0.24
David Grierson	0.25	0.23	0.24
Canberra Deep Space Communication Complex	0.23	0.22	0.23
Robert Rosenthal (USAF officer)	0.22	0.20	0.21
Johann Cruyff	0.22	0.20	0.21
Peada of Mercia	0.21	0.20	0.21
Hudson Line (Metro-North)	0.21	0.21	0.21
Last Call Cleveland	0.21	0.20	0.21
Blyth Inc	0.21	0.20	0.20
Chatuchak Park	0.20	0.10	0.11

References

1. Workshop on machine learning for ie. ECAI 2000, Berlin (2000)
2. Workshop on adaptive text extraction and mining held in conjunction with the 17th International Conference on Artificial Intelligence, IJCAI 2001, Seattle (August 2001)
3. Myspace for the dudes in lab coats. The New Scientist 192(2574), 29–29 (2006)
4. Active microscopic cellular image annotation by superposable graph transduction with imbalanced labels (2008)
5. Agosti, M., Ferro, N.: Annotations: Enriching a digital library. In: Koch, T., Sølvberg, I.T. (eds.) ECDL 2003. LNCS, vol. 2769, pp. 88–100. Springer, Heidelberg (2003)
6. Agosti, M., Ferro, N.: An information service architecture for annotations. In: Pre-proceedings of the 6th Thematic Workshop of the EU Network of Excellence DELOS, p. 115 (2004)
7. Agosti, M., Ferro, N.: Annotations as context for searching documents. In: Crestani, F., Ruthven, I. (eds.) CoLIS 2005. LNCS, vol. 3507, pp. 155–170. Springer, Heidelberg (2005)
8. Agosti, M., Ferro, N.: A system architecture as a support to a flexible annotation service. In: Türker, C., Agosti, M., Schek, H.-J. (eds.) Peer-to-Peer, Grid, and Service-Orientation in Digital Library Architectures. LNCS, vol. 3664, pp. 147–166. Springer, Heidelberg (2005)
9. Agosti, M., Ferro, N.: Search strategies for finding annotations and annotated documents: The FAST service. In: Larsen, H.L., Pasi, G., Ortiz-Arroyo, D., Andreasen, T., Christiansen, H. (eds.) FQAS 2006. LNCS (LNAI), vol. 4027, pp. 270–281. Springer, Heidelberg (2006)
10. Von Ahn, L., Blum, M., Hopper, N., Langford, J.: CAPTCHA: Using hard AI problems for security. In: Biham, E. (ed.) EUROCRYPT 2003. LNCS, vol. 2656, pp. 294–311. Springer, Heidelberg (2003)
11. Von Ahn, L., Blum, M., Langford, J.: Telling humans and computers apart automatically. ACM Commun. 47(2), 56–60 (2004)
12. Von Ahn, L., Dabbish, L.: Labeling images with a computer game. In: Proceedings of the SIGCHI Conference on Human Factors in Computing Systems, pp. 319–326. ACM, New York (2004)
13. Von Ahn, L., Liu, R., Blum, M.: Peekaboom: a game for locating objects in images. In: Proceedings of the SIGCHI Conference on Human Factors in Computing Systems, p. 64. ACM, New York (2006)

14. Alani, H., Dasmahapatra, S., Gibbins, N., Glaser, H., Harris, S., Kalfoglou, Y., Hara, K.O., Shadbolt, N.: Managing reference: Ensuring referential integrity of ontologies for the semantic web. In: Gómez-Pérez, A., Benjamins, V.R. (eds.) EKAW 2002. LNCS (LNAI), vol. 2473, pp. 317–334. Springer, Heidelberg (2002)
15. Albrechtsen, H., Andersen, H., Cleal, B.: Work centered evaluation of collaborative systems - the collate experience. In: WETICE 2004: Proceedings of the 13th IEEE International Workshops on Enabling Technologies: Infrastructure for Collaborative Enterprises, pp. 167–172. IEEE Computer Society Press, Washington, DC (2004)
16. Allen, D., Wilson, T.: Information overload: context and causes. New Review of Information Behaviour Research 4(1), 31–44 (2003)
17. Amatriain, X., Massaguer, J., Garcia, D., Mosquera, I.: The clam annotator a cross-platform audio descriptors editing tool. In: ISMIR 2005: Proceedings of 6th International Conference on Music Information Retrieval, London, UK, September 11-15 (2005)
18. Amitay, E., HarEl, N., Sivan, R., Soffer, A.: Web-a-where: geotagging web content. In: SIGIR 2004: Proceedings of the 27th Annual International ACM SIGIR Conference on Research and Development in Information Retrieval, pp. 273–280. ACM, New York (2004)
19. Andrews, K., Faschingbauer, J., Gaisbauer, M., Pichler, M., Schip Inger, J.: Hyper-g: A new tool for distributed hypermedia. In: International Conference on Distributed Multimedia Systems and Applications, pp. 209–214 (1994)
20. Antoniou, G., Van Harmelen, F.: Web Ontology Language: OWL. In: International Handbooks on Information Systems, ch. 4. Springer, Heidelberg (2009)
21. Arandjelovic, O., Cipolla, R.R.: Automatic cast listing in feature-length films with anisotropic manifold space. In: CVPR 2006: Proceedings of the 2006 IEEE Computer Society Conference on Computer Vision and Pattern Recognition, pp. 1513–1520. IEEE Computer Society, Washington, DC (2006)
22. Assfalg, J., Bertini, M., Colombo, C., Bimbo, A., Nunziati, W.: Semantic annotation of soccer videos: automatic highlights identification. Comput. Vis. Image Underst. 92(2-3), 285–305 (2003)
23. Bahadori, S., Cesta, A., Iocchi, L., Leone, G., Nardi, D., Pecora, F., Rasconi, R., Scozzafava, L.: Towards ambient intelligence for the domestic care of the elderly, pp. 15–38 (2005)
24. Banko, M., Brill, E.: Scaling to very very large corpora for natural language disambiguation. In: Proceedings of the 39th Annual Meeting on Association for Computational Linguistics, p. 33 (2001)
25. Baum, L.: The wonderful wizard of Oz. Elibron Classics (2000)
26. Beni, G., Wang, J.: Swarm intelligence in cellular robotic systems. In: Proceedings of NATO Advanced Workshop on Robots and Biological Systems, NATO, Tuscany, Italy(1989)
27. Berghel, H.: Cyberspace 2000: dealing with information overload. ACM Commun. 40(2), 19–24 (1997)
28. Bergman, M.: The deep web: Surfacing hidden value. Journal of Electronic Publishing 7(1) (August 2001)
29. Berners-Lee, T., Handler, J., Lassilla, O.: The semantic web. Scientific American Magazine (May 2001)
30. Brush Bernheim, A., Bargeron, D., Grudin, J., Borning, A., Gupta, A.: Supporting interaction outside of class: anchored discussions vs. discussion boards. In: CSCL 2002: Proceedings of the Conference on Computer Support for Collaborative Learning, pp. 425–434. International Society of the Learning Sciences (2002)

31. Besmer, A., Lipford, H.: Tagged photos: concerns, perceptions, and protections. In: Proceedings of the 27th International Conference Extended Abstracts on Human Factors in Computing Systems, pp. 4585–4590. ACM, New York (2009)

32. Bonabeau, E., Dorigo, M., Theraulaz, G.: Swarm Intelligence: From Natural to Artificial Systems. Oxford University Press, Oxford (1999)

33. Bontcheva, K., Tablan, V., Maynard, D., Cunningham, H.: Evolving gate to meet new challenges in language engineering. Natural Language Engineering 10(3/4), 349–374 (2004)

34. Bottoni, P., Civica, R., Levialdi, S., Orso, L., Panizzi, E., Trinchese, R.: Madcow: a multimedia digital annotation system. In: AVI 2004: Proceedings of the Working Conference on Advanced Visual Interfaces, pp. 55–62. ACM, New York (2004)

35. Bottoni, P., Levialdi, S., Labella, A., Panizzi, E., Trinchese, R., Gigli, L.: Madcow: a visual interface for annotating web pages. In: AVI 2006: Proceedings of the Working Conference on Advanced Visual Interfaces, pp. 314–317. ACM, New York (2006)

36. Boufaden, N.: An ontology-based semantic tagger for ie system. In ACL 2003: Proceedings of the 41st Annual Meeting on Association for Computational Linguistics, pp. 7–14. Association for Computational Linguistics (2003)

37. Boughanem, M., Sabatier, P.: Management of uncertainty and imprecision in multimedia information systems: Introducing this special issue. International Journal of Uncertainty, Fuzziness and Knowledge-Based Systems 11 (2003)

38. Bozsak, E., Ehrig, M., Handschuh, S., Hotho, A., Maedche, A., Motik, B., Oberle, D., Schmitz, C., Staab, S., Stojanovic, L., Stojanovic, N., Studer, R., Stumme, G., Sure, Y., Tane, J., Volz, R., Zacharias, V.: Kaon towards a large scale semantic web, pp. 231–248 (2002)

39. Brickley, D., Guha, R.: Resource description framework (rdf) schema specification. proposed recommendation. In: World Wide Web Consortium (1999)

40. Brill, E.: A simple rule-based part of speech tagger. In: Proceedings of the Workshop on Speech and Natural Language, p. 116. Association for Computational Linguistics (1992)

41. Brin, S.: Extracting patterns and relations from the world wide web. In: Atzeni, P., Mendelzon, A.O., Mecca, G. (eds.) WebDB 1998. LNCS, vol. 1590, pp. 172–183. Springer, Heidelberg (1999)

42. Broekstra, J., Kampman, A., van Harmelen, F.: Sesame: A generic architecture for storing and querying RDF and RDF schema. In: Horrocks, I., Hendler, J. (eds.) ISWC 2002. LNCS, vol. 2342, pp. 54–68. Springer, Heidelberg (2002)

43. Bush, V.: As we think. The Atlantic Monthly (July 1945)

44. Califf, M.E.: Relational learning techniques for natural language extraction. Tech. Report AI98-276 (1998)

45. Carroll, J.: Matching RDF graphs, pp. 5–15 (2002)

46. Carroll, J., Briscoe, T., Sanfilippo, A.: Parser evaluation: a survey and a new proposal. In: Proceedings of the 1st International Conference on Language Resources and Evaluation, pp. 447–454. Citeseer (1998)

47. Chapman, S., Dingli, A., Ciravegna, F.: Armadillo: harvesting information for the semantic web. In: SIGIR 2004: Proceedings of the 27th Annual International ACM SIGIR Conference on Research and Development in Information Retrieval, pp. 598–598. ACM, New York (2004)

48. Chen, Y., Shao, J., Zhu, K.: Automatic annotation of weakly-tagged social images on flickr using latent topic discovery of multiple groups. In: Proceedings of the 2009 Workshop on Ambient Media Computing, pp. 83–88. ACM, New York (2009)

49. Chetcuti, M., Dingli, A.: Exploiting Social Networks for Image Indexing (October 2008)
50. Chirita, P., Costache, S., Nejdl, W., Handschuh, S.: P-tag: large scale automatic generation of personalized annotation tags for the web. In: WWW 2007: Proceedings of the 16th International Conference on World Wide Web, pp. 845–854. ACM, New York (2007)
51. Cimiano, P., Handschuh, S., Staab, S.: Towards the self-annotating web. In: WWW 2004: Proceedings of the 13th International Conference on World Wide Web, pp. 462–471. ACM, New York (2004)
52. Ciravegna, F.: Adaptive information extraction from text by rule induction and generalisation. In: Proceedings of 17th International Joint Conference on Artificial Intelligence, IJCAI (2001)
53. Ciravegna, F., Chapman, S., Dingli, A., Wilks, Y.: Learning to harvest information for the semantic web. In: Bussler, C.J., Davies, J., Fensel, D., Studer, R. (eds.) ESWS 2004. LNCS, vol. 3053, pp. 312–326. Springer, Heidelberg (2004)
54. Ciravegna, F., Dingli, A., Guthrie, D., Wilks, Y.: Integrating information to bootstrap information extraction from web sites. In: Proceedings of the IJCAI Workshop on Information Integration on the Web, pp. 9–14. Citeseer (2003)
55. Ciravegna, F., Dingli, A., Petrelli, D.: Active Document Enrichment using Adaptive Information Extraction from Text. In: Horrocks, I., Hendler, J. (eds.) ISWC 2002. LNCS, vol. 2342, Springer, Heidelberg (2002)
56. Ciravegna, F., Dingli, A., Petrelli, D., Wilks, Y.: Timely and non-intrusive active document annotation via adaptive information extraction. In: Workshop Semantic Authoring Annotation and Knowledge Management (European Conf. Artificial Intelligence), Citeseer (2002)
57. Ciravegna, F., Dingli, A., Petrelli, D., Wilks, Y.: User-system cooperation in document annotation based on information extraction. In: Gómez-Pérez, A., Benjamins, V.R. (eds.) EKAW 2002. LNCS (LNAI), vol. 2473, p. 122. Springer, Heidelberg (2002)
58. Ciravegna, F., Dingli, A., Wilks, Y., Petrelli, D.: Amilcare: adaptive information extraction for document annotation. In: Proceedings of the 25th Annual International ACM SIGIR Conference on Research and Development in Information Retrieval, p. 368. ACM, New York (2002)
59. Ciravegna, F., Wilks, Y.: Designing adaptive information extraction for the semantic web in amilcare. In: Annotation for the Semantic Web. Series Frontiers in Artificial Intelligence and Applications, Artificial Intelligence and Applications. IOS Press, Amsterdam (2003)
60. Clarke, C., Cormack, G., Lynam, T.: Exploiting redundancy in question answering. In: Proceedings of the 24th Annual International ACM SIGIR Conference on Research and Development in Information Retrieval, p. 365. ACM, New York (2001)
61. Coffman, K., Odlyzko, A.: The size and growth rate of the internet. Technical report (1999)
62. Coffman, K., Odlyzko, A.: Internet growth: is there a "moore's law" for data traffic?, pp. 47–93 (2002)
63. Cohen, I., Medioni, G.: Detection and tracking of objects in airborne video imagery. Technical report. In: Proc. Workshop on Interpretation of Visual Motion (1998)
64. Cole, J., Suman, M., Schramm, P., Lunn, R., Aquino, J.: The ucla internet report surveying the digital future year three. Technical report, UCLA Center for Communication Policy (February 2003)
65. Cornish, D., Dukette, D.: The Essential 20: Twenty Components of an Excellent Health Care Team. RoseDog Books (October 2009)

66. Cunningham, H., Maynard, D., Bontcheva, K., Tablan, V.: Gate: an architecture for development of robust hlt applications. Recent Advanced in Language Processing, 168–175 (2002)
67. Cunningham, H., Maynard, D., Tablan, V.: Jape: a java Annotation Patterns Engine (1999)
68. Dingli, A., Seychell, D., Kallai, T.: igital information navigation and orientation system for smart cities (dinos). In: First European Satellite Navigation Conference, GNSS (October 2010)
69. Daniel, R., Mealling, M.: Urc scenarios and requirements. Draft, Internet Engineering Task Force (November 1994)
70. David, D., Aberdeen, J., Hirschman, L., Kozierok, R., Robinson, P., Vilain, M.: Mixed-initiative development of language processing systems. In: Fifth Conference on Applied Natural Language Processing, pp. 348–355 (April 1997)
71. Davis, J., Huttenlocher, D.: Shared annotation for cooperative learning. In: CSCL 1995: The First International Conference on Computer Support for Collaborative Learning, pp. 84–88. Lawrence Erlbaum, Mahwah (1995)
72. Dawkins, R.: The blind watchmaker. Penguin Harmondsworth (1991)
73. De Roure, D., Goble, C., Stevens, R.: The design and realisation of the virtual research environment for social sharing of workflows. Future Generation Computer Systems 25(5), 561–567 (2009)
74. Dempsey, T.: Delphic Oracle: Its Early History, Influence and Fall. Kessinger Publishing (2003)
75. Oxford Dictionaries. Concise Oxford English Dictionary, 11th edn., p. 53. Oxford University Press, Oxford (August 2008)
76. Dingli, A., Abela, C.: A pervasive assistant for nursing and doctoral staff. In: Proceedings of the Poster Track of the 18th European Conference on Artificial Intelligence (July 2008)
77. Dingli, A., Abela, C.: Pervasive nursing and doctoral assistant (pinata). In: Bechhofer, S., Hauswirth, M., Hoffmann, J., Koubarakis, M. (eds.) ESWC 2008. LNCS, vol. 5021, Springer, Heidelberg (2008)
78. Dingli, A., Ciravegna, F., Wilks, Y.: Automatic semantic annotation using unsupervised information extraction and integration. In: Proceedings of SemAnnot 2003 Workshop, Citeseer (2003)
79. Dingli, A., Seychell, D.: Virtual mobile city guide. In: Proc. of 9th World Conference on Mobile and Contextual Learning, mLearn (October 2010)
80. Dix, A., Finlay, J., Abowd, G., Beale, R.: Human-Computer Interaction, 3rd edn. Prentice-Hall, Englewood Cliffs (2003)
81. Domingue, J.B., Lanzoni, M., Motta, E., Vargas-Vera, M., Ciravegna, F.: MnM: Ontology driven semi-automatic and automatic support for semantic markup. In: Gómez-Pérez, A., Benjamins, V.R. (eds.) EKAW 2002. LNCS (LNAI), vol. 2473, p. 379. Springer, Heidelberg (2002)
82. Doswell, J.: Augmented learning: Context-aware Mobile Augmented Reality Architecture for Learning, pp. 1182–1183 (2006)
83. Douglas, T., Barrington, L., Gert, L., Mehrdad, Y.: Combining audio content and social context for semantic music discovery. In: SIGIR 2009: Proceedings of the 32nd International ACM SIGIR Conference on Research and Development in Information Retrieval, pp. 387–394. ACM, New York (2009)
84. Edwards, P., Johnson, L., Hawkesand, D., Fenlon, M., Strong, A., Gleeson, M.: Clinical Experience and Perception in Stereo Augmented Reality Surgical Navigation, pp. 369–376 (2004)

85. Enfield, N.: The Anatomy of Meaning: Speech, Gesture, and Composite Utterances. Cambridge University Press, Cambridge (2009)

86. Erasmus. Literary and Educational Writings, Volume 1 Antibarbari Parabolae. Volume 2 De copia De ratione studii (Collected Works of Erasmus), volume 23-24. University of Toronto Press (December 1978)

87. Etzioni, O., Banko, M., Soderland, S., Weld, D.: Open information extraction from the web. Communications of the ACM 51(12), 68–74 (2008)

88. Etzioni, O., Cafarella, M., Downey, D., Kok, S., Popescu, A., Shaked, T., Soderland, S., Weld, D., Yates, A.: Web-scale information extraction in knowitall(preliminary results). In: Proceedings of the 13th International Conference on World Wide Web, pp. 100–110. ACM, New York (2004)

89. Everingham, M., Sivic, J., Zisserman, A.: Hello! my name is. buffy automatic naming of characters in tv video. In: BMVC (2006)

90. Feiner, S., MacIntyre, B., Höllerer, T., Webster, A.: A touring machine: Prototyping 3d mobile augmented reality systems for exploring the urban environment. Personal and Ubiquitous Computing 1(4), 208–217 (1997)

91. Fensel, D., Hendler, J., Lieberman, H., Wahlster, W. (eds.): Spinning the Semantic Web: Bringing the World Wide Web to Its Full Potential. paperback edition. The MIT Press, Cambridge (1995)

92. Fensel, D., Horrocks, I., Harmelen, F., McGuinness, D., Patel-Schneider, P.: Oil: Ontology infrastructure to enable the semantic web. IEEE Intelligent Systems 16, 200–201 (2001)

93. Fiorentino, M., de Amicis, R., Monno, G., Stork, A.: Spacedesign: A mixed reality workspace for aesthetic industrial design. In: Proceedings of the 1st International Symposium on Mixed and Augmented Reality, IEEE Computer Society Press, Washington, DC (2002)

94. Fisher, D., Soderland, S., Feng, F., Lehnert, W.: Description of the UMass system as used for MUC-6. In: Proceedings of the 6th Conference on Message Understanding, pp. 127–140. Association for Computational Linguistics Morristown, NJ (1995)

95. Fitzgibbon, A.W., Zisserman, A.: On affine invariant clustering and automatic cast listing in movies. In: Heyden, A., Sparr, G., Nielsen, M., Johansen, P. (eds.) ECCV 2002. LNCS, vol. 2352, pp. 304–320. Springer, Heidelberg (2002)

96. Freitag, D.: Information extraction from html: Application of a general learning approach. In: Proceedings of the Fifteenth Conference on Artificial Intelligence AAAI (1998)

97. Freitag, D., Kushmerick, N.: Boosted wrapper induction. In: Proceedings Of The National Conference On Artificial Intelligence, pp. 577–583. AAAI Press/ MIT Press, Menlo Park, CA, Cambridge, MA, London (2000)

98. Frommholz, I., Brocks, H., Thiel, U., Neuhold, E.J., Iannone, L., Semeraro, G., Berardi, M., Ceci, M.: Document-centered collaboration for scholars in the humanities – the COLLATE system. In: Koch, T., Sølvberg, I.T. (eds.) ECDL 2003. LNCS, vol. 2769, pp. 434–445. Springer, Heidelberg (2003)

99. Geurts, J., Ossenbruggen, J., Hardman, L.: Requirements for practical multimedia annotation. In: Workshop on Multimedia and the Semantic Web, pp. 4–11 (2005)

100. Glatard, T., Montagnat, J., Magnin, I.: Texture based medical image indexing and retrieval: application to cardiac imaging. In: MIR 2004: Proceedings of the 6th ACM SIGMM International Workshop on Multimedia Information Retrieval, pp. 135–142. ACM, New York (2004)

101. Goble, C., De Roure, D.: Curating scientific web services and workflow. EDUCAUSE Review 43(5) (2008)

102. Goble, C.A., De Roure, D.C.: MYexperiment: social networking for workflow-using e-scientists. In: Proceedings of the 2nd Workshop on Workflows in Support of Large-Scale Science, WORKS 2007. ACM, New York (2007)
103. Godwin, P.: Information literacy and web 2. 0: is it just hype? Program: Electronic Library and Information Systems 43(3), 264–274 (2009)
104. Goldfarb, C.: The roots of sgml – a personal recollection (1996)
105. Gospodnetic, O., Hatcher, E.: Lucene in action: a guide to the Java search engine. Manning, Greenwich (2005)
106. Greenfield, A.: Everyware: The Dawning Age of Ubiquitous Computing, 1st edn. New Rides Publishing, Indianapolis (2006)
107. Groza, T., Handschuh, S., Moeller, K., Grimnes, G., Sauermann, L., Minack, E., Mesnage, C., Jazayeri, M., Reif, G., Gudjonsdottir, R.: The NEPOMUK project-on the way to the social semantic desktop. In: Proceedings of I-Semantics, vol. 7, pp. 201–211 (2007)
108. Gruber, T.: A translation approach to portable ontology specifications. Knowledge Acquisition 5(2), 199–220 (1993)
109. Gulli, A., Signorini, A.: The indexable web is more than 11.5 billion pages. In: WWW 2005: Special Interest Tracks and Posters of the 14th International Conference on World Wide Web, pp. 902–903. ACM Press, New York (2005)
110. Guven, S., Oda, O., Podlaseck, M., Stavropoulos, H., Kolluri, S., Pingali, G.: Social mobile augmented reality for retail. In: IEEE International Conference on Pervasive Computing and Communications, vol. 0, pp. 1–3 (2009)
111. Hakkarainen, M., Woodward, C., Billinghurst, M.: Augmented assembly using a mobile phone, pp. 167–168 (2008)
112. Halasz, F.: Reflections on notecards: seven issues for the next generation of hypermedia systems. In: HYPERTEXT 1987: Proceedings of the ACM Conference on Hypertext, pp. 345–365. ACM, New York (1987)
113. Halpin, H., Robu, V., Shepherd, H.: The complex dynamics of collaborative tagging. In: WWW 2007: Proceedings of the 16th International Conference on World Wide Web, pp. 211–220. ACM, New York (2007)
114. Handschuh, S., Staab, S.: Authoring and annotation of web pages in cream. In: WWW 2002: Proceedings of the 11th International Conference on World Wide Web, pp. 462–473. ACM, New York (2002)
115. Handschuh, S., Staab, S., Ciravegna, F.: S-CREAM – semi-automatic cREAtion of metadata. In: Gómez-Pérez, A., Benjamins, V.R. (eds.) EKAW 2002. LNCS (LNAI), vol. 2473, p. 358. Springer, Heidelberg (2002)
116. Handschuh, S., Staab, S., Ciravegna, F.: S-CREAM – semi-automatic cREAtion of metadata. In: Gómez-Pérez, A., Benjamins, V.R. (eds.) EKAW 2002. LNCS (LNAI), vol. 2473, pp. 358–372. Springer, Heidelberg (2002)
117. Handschuh, S., Staab, S., Studer, R.: Leveraging metadata creation for the semantic web with CREAM. In: Günter, A., Kruse, R., Neumann, B. (eds.) KI 2003. LNCS (LNAI), vol. 2821, pp. 19–33. Springer, Heidelberg (2003)
118. Hayes, J., Gutierrez, C.: Bipartite graphs as intermediate model for rdf. pp. 47–61 (2004)
119. Hearst, M., Rosner, D.: Tag clouds: Data analysis tool or social signaller? In: HICSS 2008: Proceedings of the Proceedings of the 41st Annual Hawaii International Conference on System Sciences. IEEE Computer Society, Los Alamitos (2008)

120. Herrera, P., Celma, O., Massaguer, J., Cano, P., Gómez, E., Gouyon, F., Koppenberger, M.: Mucosa: A music content semantic annotator. In: ISMIR 2005: Proceedings of 6th International Conference on Music Information Retrieval, London, UK, September 11-15, pp. 77–83 (2005)

121. Heymann, P., Koutrika, G., Molina, H.: Can social bookmarking improve web search? In: WSDM 2008: Proceedings of the International Conference on Web Search and Web Data Mining, pp. 195–206. ACM, New York (2008)

122. Himma, K.: The concept of information overload: A preliminary step in understanding the nature of a harmful information-related condition. Ethics and Information Technology (2007)

123. Ho, C., Chang, T., Lee, J., Hsu, J., Chen, K.: Kisskissban: a competitive human computation game for image annotation. In: HCOMP 2009: Proceedings of the ACM SIGKDD Workshop on Human Computation, pp. 11–14. ACM, New York (2009)

124. Hollink, L., Nguyen, G., Schreiber, G., Wielemaker, J., Wielinga, B., Worring, M.: Adding spatial semantics to image annotations. In: Proc. of the 5th Int'l. Workshop on Knowledge Markup and Semantic Annotation (2004)

125. Horrocks, I.: DAML+OIL: A reason-able web ontology language. In: Jensen, C.S., Jeffery, K., Pokorný, J., Šaltenis, S., Hwang, J., Böhm, K., Jarke, M. (eds.) EDBT 2002. LNCS, vol. 2287, pp. 2–13. Springer, Heidelberg (2002)

126. Jackson, D.: Scalable vector graphics (svg): the world wide web consortium's recommendation for high quality web graphics. In: SIGGRAPH 2002: ACM SIGGRAPH 2002 Conference Abstracts and Applications, pp. 319–319. ACM, New York (2002)

127. Jackson, H.: Marginalia: Readers Writing in Books. Yale University Press, New Haven (2009)

128. Jahnke, I., Koch, M.: Web 2.0 goes academia: does web 2.0 make a difference? International Journal of Web Based Communities 5(4), 484–500 (2009)

129. Janardanan, V., Adithan, M., Radhakrishnan, P.: Collaborative product structure management for assembly modeling. Computer Industry 59, 820–832 (2008)

130. Jang, C., Yoon, T., Cho, H.: A smart clustering algorithm for photo set obtained from multiple digital cameras. In: Proceedings of the 2009 ACM Symposium on Applied Computing, pp. 1784–1791. ACM, New York (2009)

131. Jansen, B., Spink, A.: How are we searching the world wide web? a comparison of nine search engine transaction logs. Information Processing & Management 42(1), 248–263 (2006)

132. Jianping, F., Yuli, G., Hangzai, L., Guangyou, X.: Automatic image annotation by using concept-sensitive salient objects for image content representation. In: SIGIR 2004: Proceedings of the 27th Annual International ACM SIGIR Conference on Research and Development in Information Retrieval, pp. 361–368. ACM, New York (2004)

133. Johnson, S.: Emergence: The Connected Lives of Ants, Brains, Cities, and Software. Scribner (2002)

134. Ka-Ping, Y.: Critlink: Better hyperlinks for the www. In: Hypertext 1998 (June 1998)

135. Kahan, J., Koivunen, M.: Annotea: an open rdf infrastructure for shared web annotations. In: Proceedings of the 10th International World Wide Web Conference, pp. 623–632 (2001)

136. Kersting, O., Dollner, J.: Interactive 3d visualization of vector data in gis. In: GIS 2002: Proceedings of the 10th ACM International Symposium on Advances in Geographic Information Systems, pp. 107–112. ACM, New York (2002)

137. Kim, J., Ohta, T., Tateisi, Y., Tsujii, J.: Genia corpus–semantically annotated corpus for bio-textmining. Bioinformatics 19 (suppl.1) (2003)

138. Kiryakov, A., Popov, B., Terziev, I., Manov, D., Ognyanoff, D.: Semantic annotation, indexing, and retrieval. Web Semantics: Science, Services and Agents on the World Wide Web 2(1), 49–79 (2004)
139. Klein, M., Visser, U.: Guest editors' introduction: Semantic web challenge. IEEE Intelligent Systems, 31–33 (2004)
140. Kleinberger, T., Becker, M., Ras, E., Holzinger, A., Müller, P.: Ambient intelligence in assisted living: Enable elderly people to handle future interfaces. In: Stephanidis, C. (ed.) UAHCI 2007 (Part II). LNCS, vol. 4555, pp. 103–112. Springer, Heidelberg (2007)
141. Koleva, B., Benford, S., Greenhalgh, C.: The properties of mixed reality boundaries. In: Proceedings of ECSCW 1999, pp. 119–137. Kluwer Academic Publishers, Dordrecht (1999)
142. Kumar, A.: Third voice trails off.... (April 2001), www.wired.com
143. Lampson, B.: Personal distributed computing: the alto and ethernet software. In: Proceedings of the ACM Conference on The history of personal workstations, pp. 101–131. ACM, New York (1986)
144. Laptev, I., Marszalek, M., Schmid, C., Rozenfeld, B.: Learning realistic human actions from movies. In: IEEE Conference on Computer Vision and Pattern Recognition, CVPR 2008, pp. 1–8 (2008)
145. Law, E., Von Ahn, L.: Input-agreement: a new mechanism for collecting data using human computation games. In: Proceedings of the 27th International Conference on Human Factors in Computing Systems, pp. 1197–1206. ACM, New York (2009)
146. Law, E., Von Ahn, L., Dannenberg, R., Crawford, M.: Tagatune: A game for music and sound annotation. In: International Conference on Music Information Retrieval (ISMIR 2007), pp. 361–364 (2003)
147. Lawrence, S., Giles, L.: Accessibility of information on the web. Nature 400(6740), 107 (1999)
148. Lee, S., Won, D., McLeod, D.: Tag-geotag correlation in social networks. In: SSM 2008: Proceeding of the 2008 ACM Workshop on Search in Social Media, pp. 59–66. ACM, New York (2008)
149. Lempel, R., Soffer, A.: PicASHOW: Pictorial authority search by hyperlinks on the web. In: Proceedings of the 10th International Conference on World Wide Web, p. 448. ACM, New York (2001)
150. Levitt, S., Dubner, S.: Freakonomics: A Rogue Economist Explores the Hidden Side of Everything. Harper Perennial (2009)
151. Light, M., Mann, G., Riloff, E., Breck, E.: Analyses for elucidating current question answering technology. Natural Language Engineering 7(04), 325–342 (2002)
152. Kushmerick, N., Califf, M.E., Freitag, D., Muslea, I.: In: Workshop on machine learning for information extraction, AAAI 1999, Orlando, Florida (July 1999)
153. Margolis, M., Resnick, D.: Third voice: Vox populi vox dei? First Monday 4(10) (October 1999)
154. Masahiro, A., Yukihiko, K., Takuya, N., Yasuhisa, N.: Development of a machine learnable discourse tagging tool. In: Proceedings of the Second SIGdial Workshop on Discourse and Dialogue, pp. 1–6. Association for Computational Linguistics, Morristown (2001)
155. Maynard, D., Cunningham, H., Bontcheva, K., Dimitrov, M.: Adapting a robust multigenre NE system for automatic content extraction. In: Scott, D. (ed.) AIMSA 2002. LNCS (LNAI), vol. 2443, pp. 264–273. Springer, Heidelberg (2002)
156. Mcafee, A.P.: enterprise 2.0: The dawn of emergent collaboration. MIT Sloan Management Review 47(3), 21–28 (2006)

157. McGonigal, J.: Reality is broken. game designers must fix it. In: TED 2010, California (February 2010)
158. Medwin, H.: Sounds in the Sea: From Ocean Acoustics to Acoustical Oceanography, 4th edn. Cambridge University Press, Cambridge (2005)
159. Midgley, T.: Discourse chunking: a tool in dialogue act tagging. In: ACL 2003: Proceedings of the 41st Annual Meeting on Association for Computational Linguistics, pp. 58–63. Association for Computational Linguistics, Morristown (2003)
160. Mikheev, A., Moens, M., Grover, C.: Named entity recognition without gazetteers. In: Proceedings of the Ninth Conference on European Chapter of the Association for Computational Linguistics, pp. 1–8. Association for Computational Linguistics Morristown, NJ (1999)
161. Mistry, P.: The thrilling potential of sixthsense technology. In: TEDIndia. Technology, Entertainment, Design (2009)
162. Mistry, P., Maes, P.: Sixthsense: a wearable gestural interface. In: International Conference on Computer Graphics and Interactive Techniques. ACM, New York (2009)
163. Mistry, P., Maes, P., Chang, L.: Wuw - wear ur world: a wearable gestural interface. In: CHI EA 2009: Proceedings of the 27th International Conference Extended Abstracts on Human Factors in Computing Systems, pp. 4111–4116. ACM, New York (2009)
164. Mitchell, M.: An introduction to genetic algorithms. The MIT Press, Cambridge (1998)
165. Mitchell, T.: Machine learning. In: WCB, p. 368. Mac Graw Hill, New York (1997)
166. Mitchell, T.: Extracting targeted data from the web. In: KDD 2001: Proceedings of the seventh ACM SIGKDD International Conference on Knowledge Discovery and Data Mining, pp. 3–3. ACM, New York (2001)
167. Miyashita, T., Meier, P., Tachikawa, T., Orlic, S., Eble, T., Scholz, V., Gapel, A., Gerl, O., Arnaudov, S., Lieberknecht, S.: An augmented reality museum guide. In: ISMAR 2008: Proceedings of the 7th IEEE/ACM International Symposium on Mixed and Augmented Reality, pp. 103–106. IEEE Computer Society, Los Alamitos (2008)
168. Myers, B.: A brief history of human-computer interaction technology. Interactions 5(2), 44–54 (1998)
169. Nelson, T.: Computer Lib/Dream Machines. Microsoft, paperback edition (October 1987)
170. Nelson, T.: The unfinished revolution and xanadu. ACM Comput. Surv. 37 (1999)
171. Newman, D.R., Bechhofer, S., DeRoure, D.: myexperiment: An ontology for e-research (October 26, 2009)
172. US Department of Commerce. A nation online: How americans are expanding their use of the internet. National Telecommunications and Information Administration (2002)
173. Olsen, S.: Ibm sets out to make sense of the web. CNETNews.com (2004)
174. O'Reilly, T.: What is web 2.0? design patterns and business models for the next generation of software (September 2005), www.oreilly.com
175. Ossenbruggen, J., Nack, F., Hardman, L.: That obscure object of desire: Multimedia metadata on the web (part i). IEEE Multimedia 12, 54–63 (2004)
176. Page, L., Brin, S., Motwani, R., Winograd, T.: The pagerank citation ranking: Bringing order to the web. Technical report, Stanford Digital Library Technologies Project (1998)
177. Page, S.: The Difference: How the Power of Diversity Creates Better Groups, Firms, Schools, and Societies. Princeton University Press, Princeton (2008)
178. Parameswaran, M., Susarla, A., Whinston, A.: P2p networking: An information-sharing alternative. Computer 34, 31–38 (2001)
179. Park, M., Kang, B., Jin, S., Luo, S.: Computer aided diagnosis system of medical images using incremental learning method. Expert Syst. Appl. 36, 7242–7251 (2009)
180. Perkins, A., Perkins, M.: The Internet Bubble. HarperBusiness (September 2001)

181. Popov, B., Kiryakov, A., Kirilov, A., Manov, D., Ognyanoff, D., Goranov, M.: KIM – semantic annotation platform. In: Fensel, D., Sycara, K., Mylopoulos, J. (eds.) ISWC 2003. LNCS, vol. 2870, pp. 834–849. Springer, Heidelberg (2003)

182. Popov, B., Kiryakov, A., Manov, D., Kirilov, A., Ognyanoff, D., Goranov, M.: Towards semantic web information extraction. In: Proc. Workshop on Human Language Technology for the Semantic Web and Web Services, Citeseer (2003)

183. Proust, M.: On Reading. paperback edition. Hesperus Press (January 2010)

184. Rahman, S.: Multimedia Technologies: Concepts, Methodologies, Tools, and Applications. Information Science Reference (June 2008)

185. Rak, R., Kurgan, L., Reformat, M.: xgenia: A comprehensive owl ontology based on the genia corpus. Bioinformation 1(9), 360–362 (2007)

186. Raymond, E.: The Cathedral and the Bazaar: Musings on Linux and Open Source by an Accidental Revolutionary. O'Reilly, Sebastopol (1999)

187. Riva, G.: Ambient intelligence in health care. Cyber. Psychology & Behavior 6(3), 295–300 (2003)

188. De Roure, D., Goble, C.: Software design for empowering scientists. IEEE Software 26(1), 88–95 (2009)

189. Rova, A., Mori, G., Dill, L.: One fish, two fish, butterfish, trumpeter: Recognizing fish in underwater video. In: IAPR Conference on Machine Vision Applications (2007)

190. Röscheisen, M., Mogensen, C., Winograd, T.: Shared web annotations as a platform for third-party value-added, information providers: Architecture, protocols, and usage examples. Technical report, Stanford University (1994)

191. Russell, B., Torralba, A., Murphy, K., Freeman, W.: LabelMe: a database and web-based tool for image annotation. International Journal of Computer Vision 77(1), 157–173 (2008)

192. Russell, B.C., Torralba, A.: Building a database of 3d scenes from user annotations. In: IEEE Conference on Computer Vision and Pattern Recognition. IEEE, Los Alamitos (2009)

193. Ryo, K., Yuko, O.: Identification of artifacts in scenery images using color and line information by rbf network. In: IJCNN 2009: Proceedings of the 2009 International Joint Conference on Neural Networks, pp. 445–450. IEEE Press, Piscataway (2009)

194. Schaffer, S.: Enlightened automata. The Sciences in Enlightened Europe, 126–165 (1999)

195. Schmitz, B., Quantz, J.: Dialogue acts in automatic dialogue interpreting. In: Proceedings of the Sixth International Conference on Theoretical and Methodological Issues in Machine Translation, pp. 33–47 (1995)

196. Shanteau, J.: Why Do Experts Disagree? Linking expertise and naturalistic decision making, 229 (2001)

197. Shen, Y., Ong, S., Nee, A.: Product information visualization and augmentation in collaborative design. Comput. Aided Des. 40, 963–974 (2008)

198. Shenghua, B., Guirong, X., Xiaoyuan, W., Yong, Y., Ben, F., Zhong, S.: Optimizing web search using social annotations. In: WWW 2007: Proceedings of the 16th international conference on World Wide Web, pp. 501–510. ACM Press, New York (2007)

199. Shih-Fu, C., Wei-Ying, M., Smeulders, A.: Recent advances and challenges of semantic image/video search. In: IEEE International Conference Acoustics, Speech and Signal Processing. IEEE, Los Alamitos (2007)

200. Shirky, C.: Ontology is overrated: Categories, links, and tags. Clay Shirky Writings About the Internet (2005)

201. SIGHCI. A Study of the Effects of Online Advertising: A Focus on Pop-Up and In-Line Ads (2004)

202. Song, H., Guimbretière, F., Lipson, H.: The modelcraft framework: Capturing freehand annotations and edits to facilitate the 3d model design process using a digital pen. ACM Trans. Comput.-Hum. Interact. 16, 14:1–14:33 (2009)
203. Soter, S.: What is a planet? Scientific American Magazine 296(1), 34–41 (2007)
204. Spink, A., Jansen, B., Wolfram, D., Saracevic, T.: From e-sex to e-commerce: Web search changes. Computer 35(3), 107–109 (2002)
205. Stamatatos, E., Fakotakis, N., Kokkinakis, G.: A practical chunker for unrestricted text. In: Proceedings of Second International Conference on Natural Language Processing-NLP, June 2-4, p. 139. Springer, Heidelberg (2000)
206. Stolcke, A., Ries, K., Coccaro, N., Shriberg, E., Bates, R., Jurafsky, D., Taylor, P., Martin, R., Van, C., Meteer, E.: Dialogue act modeling for automatic tagging and recognition of conversational speech. Computational Linguistics 26, 339–373 (2000)
207. Strzalkowski, T., Wang, J., Wise, B.: A robust practical text summarization. In: AAAI Spring Symposium Technical Report SS-98-06 (1998)
208. Stytz, M., Frieder, G., Frieder, O.: Three-dimensional medical imaging: algorithms and computer systems. ACM Comput. Surv. 23, 421–499 (1991)
209. Suchanek, F., Vojnovic, M., Gunawardena, D.: Social tags: meaning and suggestions. In: CIKM 2008: Proceeding of the 17th ACM Conference on Information and Knowledge Management, pp. 223–232. ACM, New York (2008)
210. Suh, B., Bederson, B.: Semi-automatic photo annotation strategies using event based clustering and clothing based person recognition. Interact. Comput. 19(4), 524–544 (2007)
211. Suits, F., Klosowski, J., Horn, W., Lecina, G.: Simplification of surface annotations. In: Proceedings of the 11th IEEE Visualization 2000 Conference (VIS 2000). IEEE Computer Society, Los Alamitos (2000)
212. Sunstein, C.: Infotopia: How Many Minds Produce Knowledge. Oxford University Press, Oxford (2008)
213. Surowiecki, J.: The wisdom of crowds: Why the many are smarter than the few and how collective wisdom shapes business, economies, societies, and nations. Doubleday Books (2004)
214. Svab, O., Labsky, M., Svatek, V.: Rdf-based retrieval of informationextracted from web product catalogues. In: SIGIR 2004 Semantic Web Workshop. ACM, New York (2004)
215. Thiel, U., Brocks, H., Frommholz, I., Dirsch-Weigand, A., Keiper, J., Stein, A., Neuhold, E.: COLLATE - a collaboratory supporting research on historic European films. International Journal on Digital Libraries (IJDL) 4(1), 8–12 (2004)
216. Thompson, C.A., Califf, M.E., Mooney, R.J.: Active learning for natural language parsing and information extraction. In: Sixteenth International Machine Learning Conference (ICML 1999), pp. 406–414 (June 1999)
217. Turnbull, D., Barrington, L., Torres, D., Lanckriet, G.: Semantic annotation and retrieval of music and sound effects. IEEE Transactions on Audio, Speech, and Language Processing (2008)
218. Uribe, D.: LEEP: Learning Event Extraction Patterns. PhD thesis, University of Sheffield (2004)
219. Vasudevan, V., Palmer, M.: On web annotations: promises and pitfalls of current web infrastructure. volume Track2, p. 9 (1999)
220. Vivier, B., Simmons, M., Masline, S.: Annotator: an ai approach to engineering drawing annotation. In: Proceedings of the 1st International Conference on Industrial and Engineering Applications of Artificial Intelligence and Expert Systems, vol. 1, pp. 447–455. ACM, New York (1988)

221. Von Ahn, L., Maurer, B., McMillen, C., Abraham, D., Blum, M.: recaptcha: Human-based character recognition via web security measures. Science 321(5895), 1465–1468 (2008)
222. Voss, J.: Tagging, Folksonomy and Co - Renaissance of Manual Indexing (January 2007)
223. Wagner, D., Schmalstieg, D.: Making augmented reality practical on mobile phones. In: IEEE Computer Graphics and Applications, 29th edn., pp. 12–15. IEEE, Los Alamitos (2009)
224. Wang, J., Bebis, G., Miller, R.: Robust video-based surveillance by integrating target detection with tracking. In: CVPR Workshop OTCBVS (2006)
225. Waxman, S., Hatch, T.: Beyond the basics: preschool children label objects flexibly at multiple hierarchical levels. J. Child Lang. 19(1), 153–166 (1992)
226. Welty, C., Ide, N.: Using the right tools: Enhancing retrieval from marked-up documents. Computers and the Humanities 33(1-2), 59–84 (1999)
227. Whitfield, S.: Life along the Silk Road. University of California Press (August 2001)
228. Wilks, Y., Brewster, C.: Natural language processing as a foundation of the semantic web. Found. Trends Web Sci. 1(3-4), 199–327 (2009)
229. Willis, R.: An attempt to analyse the automaton chess player, of Mr. de Kempelen. to which is added, a. collection of the knight's moves over the chess board. Booth (1821)
230. Würmlin, S., Lamboray, E., Staadt, O., Gross, M.: 3d video recorder. In: Proceedings of Pacific Graphics, pp. 325–334 (2002)
231. Yan, Y., Wang, C., Zhou, A., Qian, W., Ma, L., Pan, Y.: Efficiently querying rdf data in triple stores. In: WWW 2008: Proceeding of the 17th International Conference on World Wide Web, pp. 1053–1054. ACM, New York (2008)
232. Yanbe, Y., Jatowt, A., Nakamura, S., Tanaka, K.: Can social bookmarking enhance search in the web? In: JCDL 2007: Proceedings of the 7th ACM/IEEE-CS Joint Conference on Digital Libraries, pp. 107–116. ACM, New York (2007)
233. Yilmaz, A., Javed, O., Shah, M.: Object tracking: A survey. ACM Comput. 38(4), 13 (2006)
234. Lu, Y., Smith, S.: Augmented reality e-commerce assistant system: Trying while shopping. In: Jacko, J.A. (ed.) HCI (2). LNCS, pp. 643–652. Springer, Heidelberg (2007)
235. Yoon, Y.-J., Ryu, H.-S., Lee, J.-M., Park, S.-J., Yoo, S.-J., Choi, S.-M.: An ambient display for the elderly. In: Stephanidis, C. (ed.) UAHCI 2007 (Part II). LNCS, vol. 4555, pp. 1045–1051. Springer, Heidelberg (2007)
236. Zajicek, M.: Web 2.0: hype or happiness? In: W4A 2007: Proceedings of the 2007 International Cross-Disciplinary Conference on Web accessibility (W4A), pp. 35–39. ACM Press, New York (2007)
237. Zammit, P.: Automatic annotation of tennis videos. Undergraduate Thesis (2009)
238. Zhang, N., Zhang, Y., Tang, J., Tang, J.: A tag recommendation system for folksonomy. In: King, I., Li, J.Z., Xue, G.R., Tang, J. (eds.) CIKM-SWSM, pp. 9–16. ACM, New York (2009)

Glossary

AAL A field that studies how computers can support humans in their daily life, placing emphasis on the physical situation and context of that person.

AR A field which studies the combination of real-world and computer-generated data.

Acoustical Oceanography The study of the sea (boundaries, contents, etc) by making use of underwater sounds.

Blog A contraction of the term "Web log". It is essentially a website which allows a user to read, write or edit information containing all sorts of multimedia. Posts are normally sorted in chronological order.

Bookmarklet A small program stored in a URL which is saved as a bookmark.

Browser add-ons Small programs used to customise or add new features to a browser.

Cloud computing Computing based around the Internet where all resource, software and processing power is obtained through an online connection.

Conversion page A term used in Search Engine Optimisation to refer to those website pages where the actual sales occurs. It is called a conversion page because it converts the user from a visitor to a buyer.

COP A community made up of a group of people who share a common interest.

Dataset A logical grouping of related data.

Deixis Words that refer to someone (such as **he, she**, etc) or something (such as **it**) which can only be decoded within a context.

Ellipses Partial phrases having missing text which can only be decoded by keeping track of the conversation.

Folksonomy The organisation of tags based upon classes defined by the users.

GeoTagging The process of adding geographical information such as latitude and longitude to digital media.

GIS A system capable of storing and analysing geographical information.

Homonomy Words having same syntax but different semantics.

Hyperlink A link from one electronic document to another.

Intelligent Agent A computer program capable of learning about its environment and take actions to influence it.

IM A technology that allows several people to chat simultaneously in real time.

Incidental Knowledge Elicitation A set of techniques used to elicitate knowledge from users as a byproduct of another process.

Knowledge Elicitation A set of techniques used to acquire knowledge from humans and learn from the data they produce.

Latitude A position on the Earth's surface which is located on a line in parallel with the equator.

Longitude A position on the Earth's surface which is located on a line perpendicular to the equator.

Meta Data Data used to describe other data.

Micro-blog Similar to a blog but with a restricted size, typically made up of a sentence or two.

Namespace A unique term used to reference a class of objects.

OCR A program which converts scanned images to editable text.

Ontology A formal specification of a shared conceptualisation.

Open Graph protocol A protocol which enables any web page to become an integral part of a social graph.

P2P A program capable of sharing files with other users across the internet without requiring a centralised server.

POI A specific location which someone might find useful. A POI is normally used in a GIS.

ReTweet A reposting of a Tweet.

RFID A technology which makes use of radio waves to localise RFID tags.

RSS A protocol used to publish news feeds over the Internet.

Semantics Derived from two Greek words semantikos and semaino which essentially refer to the problem of understanding or finding the meaning.

Social Bookmarking The facility to save and categorise personal bookmarks and share them with others.

Social Graph A graph which shows the social relationships between users in a social networking site.

Social Tagging The process of annotating and categorising content in collaboration with others. Also referred to as collaborative tagging, social classification and social indexing.

Synonymy Words having different syntax but same semantics.

Tag Cloud A visual representation of user tags. This representation can revolve around an element (text, etc) or even a web resource (such as a URL). The size of the tag in the cloud is a visual representation of its importance.

Triple A data structure consisting of three parts normally represented as a graph made up of two nodes and a relationship between those nodes.

Tweet A 140 character (or less) post on the popular social networking site Twitter.

URI A unique string which identifies a resource on the Internet.

VOIP A system that uses the Internet to transmit telephone calls.

Workflow A protocol defining a set of connected steps in a process.

WYSIWYG Refers to any system whose content (while it is being edited) appears similar to the actual output.

Index